MAUNA KEA

A Guide to Hawai'i's Sacred Mountain

A Guide to Hawai'i's Sacred Mountain

Leslie Lang and David A. Byrne

WATERMARK
PUBLISHING

ISBN 0-9753740-5-2

Library of Congress Control Number: 2005924519

Front cover image: University of Hawai'i astronomer Richard Wainscoat kept his camera shutter open all night to obtain this photograph of the Canada-France-Hawaii Telescope. The rotation of the earth causes stars to appear as arcs, or "star trails." Polaris (Hōkūpa'a or "fixed star" in Hawaiian), the pole star, is the short arc near the center of rotation.

Watermark Publishing
1088 Bishop Street, Suite 310
Honolulu, HI 96813
Telephone: Toll-free 1-866-900-BOOK
Web site: www.bookshawaii.net
e-mail: sales@bookshawaii.net

Design by Leo Gonzalez
Design/Production by Nancy Watanabe

Printed in Korea

Contents

Welcome to Hawai'i's Sacred Mountain—Mauna O Wākea

Visiting Mauna Kea is an adventure—one that allows us to step back in time in the realm of the gods of the Hawaiian people. In the stars, astronomers can trace the ancient history of the universe. But we Hawaiians go to Mauna Kea in search of our *mana*—divine power—in a quest to understand our ancient spiritual connections. On the mountain we can feel the close relationship between heaven and earth. People of many nations say that it is a sacred place for them also, where they experience awe and reverence as we do.

As a Hawaiian cultural practitioner and caretaker of the mountain, I am often asked how visitors should conduct themselves on Mauna Kea. I suggest that they say a silent prayer, take a gentle moment for greeting the mountain, and then walk with respect on our sacred place of worship. I share how we must *mālama* the *'āina*—take care of the land—take care of our people and preserve the culture.

We Hawaiians are fiercely proud of the accomplishments of our ancestors, who navigated the vast Pacific Ocean by the stars a thousand years before Galileo first pointed his telescope toward the heavens. Mauna Kea was a landmark for ancient navigators and is today a center for the evolving science of astronomy as we scope out

Ancient shrines still stand at this sacred place.

The lele, *or altar, at the summit symbolizes Mauna Kea's spiritual connection.*

our place in the universe.

As a Ranger on Mauna Kea, I've enjoyed working with many astronomers, who are generally people of goodwill and from whom I have learned much about the stars. But despite all of their scientific accomplishments, I do feel that much more needs to be done to bring awareness of and respect for Hawaiian culture on the mountain. Science does play an important role in people's lives, but it is not everything. A spiritual connection is just as important. This is symbolized for modern Hawaiians by the humble stone and wooden *lele,* the altar, at the summit.

I welcome you to Mauna Kea, also known as Mauna O Wākea—the mountain of the god Wākea, from whom all things Hawaiian are descended. Here you may experience and enjoy beautiful sunrises, sunsets and evening stargazing under the Northern and part of the Southern sky. But here, too, are preserved many magical wonders of the Hawaiian nation. We all need to continue to perpetuate and protect this land, as well as the legends and mythology passed down through the ages, for our own and future generations. We must all continue to be good stewards of this sacred mountain.

Mahalo nui loa,

—James (Kimo) Kealii Pihana
Mauna Kea Ranger
Hawaiian Cultural Practitioner

FOREWORD

The early Polynesian navigators were the scientists and astronomers of their day. They had a strong curiosity about what lay beyond the horizon and were ingenious in their development of the technologies necessary to explore the vast reaches of the Pacific. They created the double-hulled sailing canoes that would make the long journeys and discovered a means of finding their way about the trackless ocean. Wayfinding was accomplished primarily by knowledge of the stars' movement across the sky and of the ways latitude altered the stars' position. These skills allowed the explorers to discover and populate the vast area of the Pacific we call today the Polynesian Triangle: from Tahiti to Hawai'i to Easter Island. Hawai'i was settled as early as 300 A.D.

By the time Captain James Cook arrived in 1778, it appears that ocean voyaging had become a lost and mostly forgotten art. But Cook's arrival introduced into the local culture new navigational instruments and techniques for ocean-going vessels: telescopes and compasses and sextants, for example. By the time the missionaries arrived and began developing a written Hawaiian language, even more of the ancient seafaring knowledge had been lost. Though the Hawaiians must have had hundreds of names for stars and other features of the sky, these were never written down—the missionaries themselves probably didn't know enough to ask!—and today are mostly forgotten.

In 1874 King Kalākaua and the citizens of Honolulu were quite astounded and impressed by an astronomical expedition that came to Hawai'i all the way from England to observe a transit of Venus (the planet Venus passing in front of the sun). A compound along Fort Street in Honolulu was set aside for various telescopes with which to observe the phenomenon. Later, in his travels to the U.S., King Kalākaua visited the Lick Observatory in California and expressed his hope that one day Hawai'i should have such a telescope. Kalākaua, the "Merrie Monarch" who revived Hawaiian traditions, was also very forward looking and eager to bring Hawai'i into the modern world. Not until 1910, however, was an astronomical observatory first established in Hawai'i. It was developed for the specific purpose of viewing the predicted coming of Halley's Comet and was located in what is today Kaimukī, on a hill along Ocean View Drive. It was not a very good telescope, and by 1958 it was razed, just before the termites did it in.

Modern-day astronomy really got its start in 1957-58 with the advent of the International Geophysical Year and the need for observations of the sun during Hawai'i's daylight hours. This led to the construction of a solar observatory at Makapu'u Point on O'ahu as a part of a worldwide network of solar observations. During this period the University of Hawai'i also developed plans for a permanent world-class solar observatory on Haleakalā on Maui. A high-altitude site was a requisite for studying the faint glow of the sun's atmosphere or corona, which normally could be seen only during a total solar eclipse. This observatory, named the Mees Solar Observatory, was completed in 1961.

During the final construction phase of the Haleakalā observatory, an astronomer from the Lunar and Planetary Laboratory in Tucson, Arizona, took an interest in the astronomical potential of Hawai'i's high mountains. Dr. Gerard Kuiper, a world-renowned planetary astronomer, was interested in the possibility of establishing a planetary observatory in Hawai'i and responded to an invitation from the Hilo Chamber of Commerce to look into it. His evaluation of Haleakalā showed that it was an excellent site, but studies on Mauna Kea suggested that it was even better. A grant from Governor John Burns made it possible to bulldoze a road to the summit, and Kuiper's site testing led him to declare that Mauna Kea was probably the best site in the world for astronomical studies. The high altitude of the mountain (almost 14,000 feet), its isolation in the middle of the Pacific and its freedom from light contamination were among the factors that contributed to the mountain's perfection for astronomy.

Dr. Kuiper proceeded to apply to NASA for funds to build an observatory on Mauna Kea. NASA, however, thought that it should also consider proposals from other interested organizations, among them Harvard University and the University of Hawai'i. The outcome was that the University of Hawai'i was awarded the grant to construct an 88-inch (2.2-meter) telescope on the mountain. It was completed in 1970. The knowledge of Mauna Kea's outstanding qualities as an observation site spread like wildfire among the astronomical community, and within a few years several more telescopes were constructed on the mountain. Today there are some 13 telescopes on Mauna Kea, constituting the world's largest collection of astronomical telescopes at the world's best site for exploring the vast universe!

Legend has it that the ancient Polynesian navigators looked upon Mauna Kea as a beacon guiding their voyages of discovery. Today Mauna Kea is again a beacon, guiding modern explorers to discoveries beyond the celestial horizon.

—Walter R. Steiger
Professor Emeritus of
Physics and Astronomy,
University of Hawai'i

Introduction

On the Big Island of Hawai'i, everyone looks up to Mauna Kea, the spectacular mountain that stands sentinel over the tropical landscape. In winter, people watch for the first snowfall of the season, always an anomaly on an island in the middle of the Pacific. Visitors drive up the mountain on a road forged through lava flows and the Big Island's own moonscape: a geography so otherworldly that Apollo astronauts trained there before their first moonwalk, thinking it would simulate a trek across the lunar surface (turns out it didn't, because the moon is smoother and doesn't have Mauna Kea's peaks). Those who reach the summit area see multi-million-dollar astronomical observatories glistening in the sunlight—portals to another world, that of vast space.

Yet to some of us, Mauna Kea is so much more than this. Hawaiians have nurtured a close relationship with the mountain since long before there were moonwalks or high-tech observatories. For the ancients, Mauna Kea was a beacon that guided voyaging canoes to these uninhabited islands. In our cosmology, the mountain is an actual ancestor of the Hawaiian people. It's the dwelling place of the goddess Poli'ahu, whose profile can be seen on the mountain, if you know where to look. It is the source of the healing waters to treat the sick, and where some Hawaiians leave the *piko* (umbilical cord) of their newborns. It is, literally, the resting place of our ancestors.

The continuing development of astronomical observatories on Mauna Kea's summit region has caused considerable heated debate. Some people are concerned about the inadvertent destruction of ancient sites and natural habitats, others about pollution of previously untainted areas. As the quest for knowledge and space exploration continues, still others are upset over more observatories being constructed on the goddess Poli'ahu's *'ōpū* (belly).

Today, however, these issues are on the table, as the scientific and cultural communities work together to address concerns, find solutions and manage development thoughtfully and responsibly.

What can we do? As we continue to use the resources atop Mauna Kea to learn about the universe we live in and explore its origins, we must also respect the mountain's place in our own island universe, and its considerable con-

Many who visit Mauna Kea are touched by its serene presence.

tributions to our knowledge about the past.

If you travel up the mountain, remember to pass through without leaving your mark—or your trash. Stay on the road, and don't touch or move rocks; they might, unbeknownst to you, be part of an ancient or contemporary shrine. Respect the mountain's unique geologic and cultural history, its steady presence.

Leave Mauna Kea unscathed for the next generation and for the generations after that. Let them, too, see what our ancestors saw, and what we still see. Allow them to discover what remains to be seen.

—Leslie Lang

1 Visiting Mauna Kea

1 Visiting Mauna Kea

GETTING THERE

Mauna Kea stands in the central section of the Big Island, its summit roughly midway between Hilo and Kona. By car, the mountain is accessed via the Saddle Road, so named because it passes through the Humu'ula "saddle," a gentle valley that lies between the volcanoes Mauna Kea and Mauna Loa.

If you're starting from Hilo, take Kamehameha Avenue, which runs along the bayfront, to Waiānuenue Avenue at the north end of downtown. Stay to the left on Waiānuenue Avenue and it becomes Kaūmana Drive, and then turns into Saddle Road (Hwy. 200).

From Kona, take Highway 190 toward Waimea, and then turn right onto Saddle Road after the 7-mile marker.

If you're starting in Waimea, head toward Kona on Hwy. 190 and turn left on the (marked) Saddle Road after the 6-mile marker.

Once on Saddle Road, watch for Pu'u Hulahula at the 28-mile marker sign, where there's a hunters' check-in station and rest area. Across from it is a paved road—it's unmarked—leading up Mauna Kea to the Visitor Information Station, six miles up at the 9,200-foot (2,804-meter) elevation.

The infamous Saddle Road, about 50 miles long, is a narrow, winding road connecting Hilo and Kona, and driving it is a unique experience. The undeveloped, desolate landscape looks rather other worldly in places, especially when a heavy fog hangs over it, which is not uncommon. Views gradually change from fern-covered forests on the Hilo end of the road to fields of lava and then rolling green hills on the Waimea side.

The road was hastily built in 1942 by the military, which needed a quick route for their heavy armored vehicles to move between Hilo and Kona. Not constructed with everyday drivers (and cars) in mind, in spots the narrow, winding two-lane highway looks like one lane divided down the middle with a yellow line. One of the reasons the Saddle Road winds so much is it was deliberately constructed to make it harder for (Japanese) planes to bomb military convoys travelling it. The road, which crests at 6,500 feet (1,981 meters), dips, rises and sometimes disappears altogether in thick fog and mist. However, it is better maintained now than in the past.

If you do drive the Saddle Road, take it slowly (but pull over

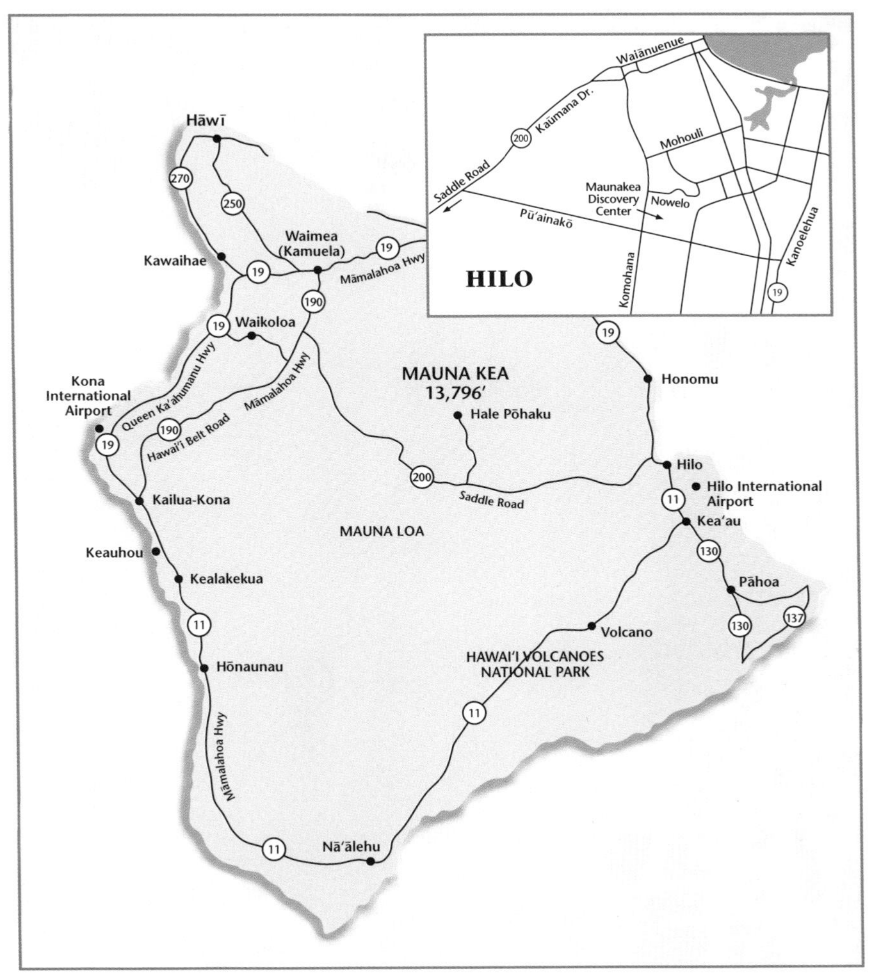

and let other cars pass if necessary). Military vehicles from the nearby Pōhakuloa Training Area (119,000 acres used for live-weapon, heavy-artillery and bomb training as well as troop maneuvers) often travel this road, so watch for convoys of tanks and armored personnel carriers, as well as the occasional low-flying military helicopter. And remember to use the restroom, fill the car with gas and stock up on

Distances and drive times to the Visitor Information Station:

From Hilo	35 miles (56 kilometers), about an hour
From Kona	70 miles (113 kilometers), 1 hour and 45 minutes to 2 hours
From Waimea	39 miles (63 kilometers), about an hour

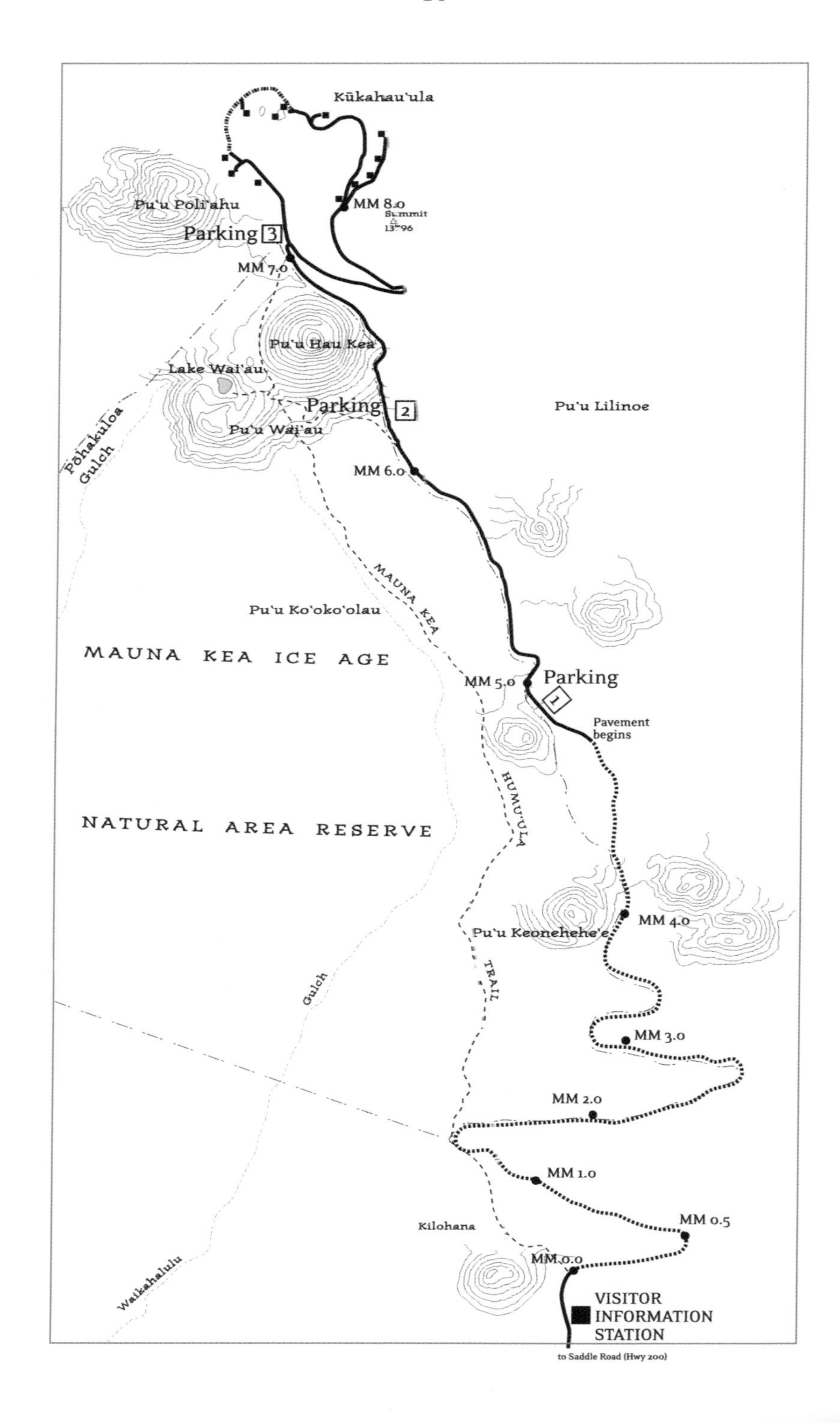
Kūkahau'ula
Pu'u Poli'ahu
MM 8.0
Summit
Parking 3
MM 7.0
Pu'u Hau Kea
Lake Wai'au
Parking 2
Pu'u Lilinoe
Pu'u Wai'au
Pōhakuloa Gulch
MM 6.0
MAUNA KEA
Pu'u Ko'oko'olau
MAUNA KEA ICE AGE
MM 5.0
Parking 1
Pavement begins
HUMU'ULA
NATURAL AREA RESERVE
MM 4.0
Pu'u Keonehehe'e
TRAIL
Gulch
MM 3.0
MM 2.0
MM 1.0
MM 0.5
Kilohana
MM 0.0
Waikahalulu
VISITOR INFORMATION STATION
to Saddle Road (Hwy 200)

food and water before leaving town, because there are no facilities on most of Saddle Road.

ALONG THE WAY

As you head from Hilo up Waiānuenue Avenue, you're near the Wailuku River, which runs to your right and is the approximate boundary where newer Mauna Loa lava flows lie alongside (more accurately, atop) older ones from Mauna Kea.

Not long after you pass Komohana Avenue, Waiānuenue Avenue splits. To proceed to Mauna Kea, take the left fork, Kaūmana Drive, which becomes Saddle Road.

To do a little sightseeing, though, it's worth staying on Waiānuenue Avenue and then turning right on Rainbow Drive, shortly after the road splits. Rainbow Drive takes you to nearby Rainbow Falls, a picturesque spot on the river with a beautiful waterfall and natural pool. Or continue up Waiānuenue Avenue and turn right on Pe'epe'e Street, which takes you to Wailuku River Park and Boiling Pots. That's another scenic spot, where the Wailuku River forms natural pool after pool that seem to bubble. At Boiling Pots, water drains between pools through underground lava tubes or fissures. You can see whirlpools where the water is funneling out of one pool, and then see it "boil" up in the next pool downstream.

Once you're back on Saddle Road, watch for the 4-mile marker. A few hundred yards beyond it you'll see Kaūmana Cave County Park on your right. The cave is the main lava tube fed by the 1881 Mauna Loa eruption, which sent lava billowing dramatically and threateningly toward Hilo.

Early visitors pose at lava tubes.

Word has it that Princess Ruth Ke'elikolani came from Honolulu to ask the volcano goddess Pele to spare Hilo town. Arriving in Hilo, she rented a hack; then she and her 25 to 30 attendants, who carried tents and two roasted pigs, proceeded, chanting, to the flow. She sent one of her people back to town for as many red silk handkerchiefs as he could round up—because the volcano goddess Pele likes red silk, she said—and he bought 30 of them at Turner's Dry Goods Store. He also picked up a quart of brandy. Princess Ruth smashed the bottle on the lava as an offering to Pele, then laid out her blankets and slept next to the flow. By morning the lava had stopped right where she offered the brandy—just a mile and a half from the town.

Kaūmana Cave sits in a county park with a small picnic area. There is easy access into the cave: a staircase, festooned at the bottom with ferns and bright impatiens, leads down to the cave opening.

The cave at Kaūmana is part of a lava tube, and its entrance is a collapsed skylight. A lava tube forms when the outside portion of a river of lava begins to cool and harden, while the inside remains liquid and flows through. When the lava drains out, it leaves behind a hollow tube. In this case, the cave roof is 20 to 25 feet (six to eight meters) thick.

Don't venture into the cave without a flashlight. This isn't the controlled environs of Disneyland: though the walking isn't difficult, you'll need a light to see where to put your feet on the sharp, uneven lava. Look for tiny lava stalactites and lava drip tracks on the upper walls and ceiling. Also hanging from the ceiling are tree roots that support some of the insects and spiders living in the cave.

Another cave—actually part of the same lava tube but on the other side of the collapsed skylight entrance—is a little more rugged and a good destination for more avid spelunkers.

After leaving Kaūmana Cave, the road climbs lava from the 1881 flow, then passes over lava from the 1855 and the 1935 and '36 flows. You'll also see some *kīpuka*, isolated islands of growth flourishing in the midst of a lava field; they're places where lava detoured around old growth without destroying it.

If you're driving Saddle Road from Kona to Mauna Kea, you'll see sand dunes where very fine sand, ground down by glaciers, blew off Mauna Kea after the glaciers melted (see "Hawai'i's Ice Age," p. 52). That's at 45 miles (72 kilometers) (there are signs marking off the miles). You can look up the mountain and see the *māmane-naio* forest where the *palila* birds live (see "Palila," p. 59). You'll be able to spot it; it's the only forest area up there. At 43.5 miles (70 kilometers) is a four-wheel-drive dirt road that goes up into the Forest Reserve.

BOMBING LAVA

Coming from Hilo, at about the 23-mile marker on the Saddle Road, you cross an *'a'ā* (rough) lava flow from 1935 and '36. This flow has an interesting history: it marks the only time anyone tried to control the flow of lava by dropping bombs.

Thomas Jaggar, director of the Hawai'i Volcanoes Observatory at the time, got the U.S. Army Air Corps to drop 500-pound bombs on the lava. They aimed their attack at the mouth of the lava tube feeding the main flow, hoping to divert the flow's path so it would not inundate Hilo. Geologist Harold Stearns wrote about the bombing:

I obtained a seat in one of the planes as an observer. Jaggar stood in the Saddle area below us with a transit to watch the operation. Most of the bombs fell too far from the channel to do any good, and a few didn't even explode, as I found out from a field inspection later. One bomb hit the tube and threw out clots of lava but did not divert the lava from its channel. The flow stopped two days later, and Jaggar believed that the bombing had stopped the flow. I am sure it was a coincidence, however, like the time, in 1881, when Princess Ruth Keelikolani tried to stop a lava flow from invading

Many have attempted to divert lava.

Hilo by throwing thirty red silk handkerchiefs and a bottle of brandy into the molten flow. Several old Hawaiians told me later that all who dropped bombs would die by fire because of arousing the wrath of Madame Pele, the goddess of volcanoes. Later, I learned that many did die by fire when their planes crashed. Maybe I did not bring down the wrath of Madame Pele because I was just an observer.

Another possible side trip en route from Hilo to the mountain is the eight-mile (13-kilometer) Pu'u'ō'ō hike. The hike starts at a marked trailhead near a small lava parking lot on the left, between Saddle Road's mile markers 22 and 23. This trail, which passes through woods and meadows as well as across barren lava, offers up a tremendous amount of bird life. It's best to start this hike in the morning, as afternoons on Saddle Road can be very foggy.

Mauna Kea State Park is at the 35-mile marker, coming from the Kona side, and has the only restrooms along Saddle Road. Cabins with cooking facilities and heaters are available, or you can stop and picnic among the native *naio* and *māmane* trees (there are also non-native eucalyptuses and pines). The park has nice views of both Mauna Kea and Mauna Loa. Call 974-6200 in advance for cabin reservations. It's not far from the Pōhakuloa Training Area, so keep in mind that your peace and quiet might occasionally be interrupted by the sounds of military maneuvers.

ALTITUDE AND SAFETY

We urge you to read this section carefully before visiting Mauna Kea and take appropriate precautions regarding your health and well-being on the mountain.

Altitude is categorized as high (8,000 to 12,000 feet or 2,438 to 3,658 meters), very high (12,000 to 18,000 feet or 3,658 to 5,486 meters), or extremely high (18,000+ feet or 5,486+ meters). Mauna Kea's summit falls into the "very high" category, and the possibility of altitude sickness should be taken seriously.

At the summit, you are above 40% of the earth's atmosphere.

To lessen your risk of altitude sickness:

Do

- Let your body adjust a bit by spending at least a half-hour (an hour is better) at the Visitor Information Station (9,200 feet or 2,804 meters) before traveling up to the summit area.
- Drink plenty of water, both before you ascend the mountain and while you're up there, to avoid dehydration.
- Eat a reasonable amount, but not so much that your digestive system is overtaxed.

Don't

- Smoke for 48 hours before going up the mountain. This will increase blood supply to your lungs and increase your lung capacity.
- Eat foods such as beans, onions and cabbage. At high altitudes, these can cause intestinal gas to expand and result in flatulence, bowel distension and pain.

Atmospheric pressure is roughly 60% what it is at sea level. This means less oxygen is available to the lungs, and shortness of breath is experienced after even moderate exertion. Not everyone's body adjusts well to the reduction in atmospheric pressure. High altitudes can also cause life-threatening conditions such as pulmonary edema (fluid in the lungs) and cerebral edema (fluid on the brain). Though most people are fine at the Visitor Information Station (9,200 feet or 2,804 meters), many visitors to the summit area (13,796 feet or 4,205 meters) get altitude sickness.

It's impossible to predict who will get altitude sickness. The condition—which is extremely unpleasant and can be serious or even life threatening—hits young and old, men and women, healthy and unfit alike. Some people just get it and others don't, and there's no way to know ahead of time who will succumb.

Children under 16, pregnant women, the severely overweight, and anyone with high blood pressure, diabetes, or heart or respiratory conditions should not venture above the Visitor Information Station. Don't plan to ascend the mountain within 24 hours of scuba diving, either, or you may suffer "the bends" (make it 48 hours if you dived below 100 feet or 30.5 meters). Don't go up the mountain if you've been drinking alcohol.

Symptoms of altitude sickness include shortness of breath, headache, dehydration, nausea, poor judgment and drowsiness, disorientation, blue lips or fingernails, vomiting, flu-like symptoms, and loss of balance and muscle coordination. Another symptom is an increased respiratory rate, which can cause hyperventilation, lightheadedness and a general feeling of body tingling. Any of these symptoms is a warning to leave the summit area immediately and head down the mountain. The only effective treatment is to descend to a lower altitude.

Take these guidelines seriously. Mauna Kea is remote, and emergency services, if needed, could take hours to reach you.

Another concern at the summit area is sunburn. At this altitude there's less atmosphere to protect you from the sun's harmful ultraviolet rays. Protect your skin! It's possible to get first- and even second-degree burns after as little as 15 minutes at the summit if you are unprepared and unprotected. Keep in mind, too, that the sun can damage your eyes, especially if it's reflected off snow. Wear sunglasses with UV protection.

WEATHER

As hard as it is to imagine when you're getting dressed in the tropical weather down by the beach, the weather on the mountain can be severe. During the winter, there is often snow at the summit, sometimes several feet of it. Temperatures usually range from about 25 to 40 degrees Fahrenheit (-4 to +4 Celsius), though wind chill and the high altitude can make it feel much colder. From April to November the weather is milder, but it still ranges from freezing to around 60 degrees Fahrenheit (0 to 16 Celsius).

Ice can form on the summit area under different conditions.

Winter weather conditions at the summit can change dramatically and quickly. Ice sometimes forms suddenly and without warning, causing dangerous conditions on the steep roads. Winds of more

It's hard to imagine this from a tropical beach.

than 100 mph can occur. Snowstorms and "white-outs" caused by fog and blowing snow sometimes reduce visibility to zero.

Deep snowdrifts and freezing fog can close roads. Keep this in mind—and should you see "Road Closed" signs, heed them. If hazardous conditions begin developing, immediately leave the summit for lower elevations.

Have we convinced you to bring appropriate cold-weather clothing? Even in the summer, bring warm jackets, gloves, long underwear, hats, scarves and any other warm clothing you can find or borrow. You'll be glad you went to the trouble.

Don't forget to take care of your skin and eyes by using sunscreen and lip protection and wearing sunglasses. Hiking boots are not necessary but are great if you have them. Otherwise, wear sturdy walking shoes and, if possible, wool socks.

Call 935-6268 for recorded updates on road conditions and closures.

The Mauna Kea Weather Center also gives detailed forecasts of weather on the mountain, as well as webcam images and more. See *http://mkwc.ifa.hawaii.edu*

LENTICULAR CLOUDS

Occasionally, objects that appear to be UFOs are spotted over Mauna Kea! While some people can't identify these amazing, flying saucer-shaped objects, there are others with a pretty good idea of what they are.

They're clouds—smooth, saucer-shaped clouds so remarkable that they become a source of conversation around town. ("Did you see that cloud this morning?!") Often a photo of the unusual cloud shows up on the front page of local papers

Lenticular clouds form when strong winds blow over high mountains.

Occasionally, clouds that look like UFOs are spotted over Mauna Kea.

the next day.

Lenticular clouds (*altocumulus lenticularus*) form when strong winds blow over high mountains such as Mauna Kea. The mountain forces winds around and over its peaks, which results in "waves" in the atmosphere downwind. When there's enough moisture in the atmosphere, lenticular clouds form where the "waves" peak. Unlike other clouds, they stay put and don't blow away with the wind. Sometimes they last quite a long time before they dissipate—hours, even—lending even more to their preternatural appearance.

Their unusual shape—like a stereotypical flying saucer, top and bottom surfaces smooth and sleek—is an incredible thing to see. Lenticular clouds occur infrequently in Hawai'i; they're more common in the western part of the mainland, such as in the Rocky Mountains. But they do occasionally occur, so keep a look-out. Seeing a lenticular cloud is something you won't soon forget.

DRIVING

You can make the trip to the Visitor Information Station in a regular car, but to continue up the mountain beyond it to the summit area you'll want to have four-wheel drive. The first four miles of the road beyond the VIS are unpaved, and then the last four are paved. The entire road is winding and steep: it rises nearly 5,000 feet (1,524 meters), with grades of as much as 15 percent. Driving is even more difficult when ice and snow are present.

In good weather it takes about a half-hour to drive from the VIS to the summit, but be careful and take your time. Use your headlights in the fog, and watch out for loose gravel. It's especially hard to see on the switchbacks for about an hour after sunrise and an hour before the sun sets, when you're driving into the glare. Take extra care at those times. It's recommended you use low gear when descending, to reduce the possibility of your brakes overheating, and do not exceed the 25-mph speed limit.

Often there's a temporary ridge built up by grading equipment at the center of the road. When crossing that ridge, be sure your car is high enough to clear it and watch for large rocks. Keep an eye out for road-maintenance equipment.

Stay on the road at all times. Driving off road is hazardous, not only to rare plants and animal habitats but also to subtle archaeological sites that are sometimes hard to identify with an untrained eye. You

When driving on the mountain, be sure to obey the safety signs.

Mauna Kea's beauty is worth all the effort.

can also do ecological damage by driving off the main roads. Creating new paths means making new channels for the mountain's melted snow runoff and rare rains, and can lead to rapid erosion.

Gasoline is not available on the mountain, so make sure you start out with a full tank. If you have to call for roadside assistance, it's going to be an expensive proposition.

The only permanent restrooms are at the Visitor Information Station (restrooms at the Keck I Telescope are available weekdays from 10 a.m. to 4 p.m., and portable ones at Parking Area #3 and near the UH 0.6-meter Telescope are always accessible); and the only public telephones are a pay phone on the Visitor Information Station patio and an emergency phone in the entrance to the University of Hawai'i 2.2-meter Telescope building (both are available night and day, every day of the year).

Alcohol is not permitted on the mountain, but do bring food and water.

Check current road conditions before you go at *http://mkwc.ifa.hawaii.edu/current/road-conditions*

TAKE A TOUR

If you don't want to drive up the mountain yourself, the following commercial tour operators have permits to conduct tours up Mauna Kea.

Contacts as of January 2005 (see *http://www.ifa.hawaii.edu/info/vis/* for an updated list):

Arnott's Lodge & Hiking Adventures
(808) 938-4870
e-mail: mahalo@arnottslodge.com
http://www.arnottslodge.com

Big Island Golf Tours
(808) 331-0733
fax: (808) 331-0833
e-mail: bigti@verizon.net
http://www.bigti.com

Hawaii Forest & Trail
(808) 331-8505
fax: (808) 331-8704
e-mail: info@hawaii-forest.com
http://www.hawaii-forest.com

Hawaiian Eyes
(808) 322-2691
fax: 808 322-2691
e-mail: masashi@gte.net

Jack's Tours
(808) 969-9507
fax: (808) 969-7681
e-mail: info@jackstours.com
http://www.jackstours.com

Meridian H.R.T.
(Japanese-language tours)
(808) 885-7484
fax: (808) 885-8871
e-mail: meridian@kona.net

Paradise Safaris
in Kona: (808) 322-2366
toll-free: (888) 322-2366
fax: (808) 325-1023
e-mail: patw@maunakea.com
http://www.maunakea.com

Roberts Hawaii
(808) 329-1688
fax: (808) 329-2631
e-mail:
casey.ballao@robertshawaii.com
http://www.robertshawaii.com

Taikobo Hawaii
(Japanese-language tours)
(808) 329-0599
fax: (808) 329-0779
e-mail: hawaii@taikobo.com
http://www.taikobo.com

Tour companies will pick you up from your hotel and provide transportation to Mauna Kea, warm clothing and a guided tour. Guided sunset tours of the summit are very popular. After a stop at the Visitor Information Station to look around and acclimate, a sunset tour proceeds to the summit area. After sunset, each tour company conducts a private stargazing program that lasts from one to three hours. Before descending the mountain, each group again stops at the Visitor Information Station for a last visit to the bookstore and to use the facilities.

Many major observatories are clustered at the summit area.

Daytime tours, nature hikes and bird-watching opportunities are also available. On certain days, tour groups can go into some of the observatories and see the telescopes. Contact the tour companies for specific information.

In general, visitors cannot go inside the observatories nor look through their telescopes. All the Mauna Kea observatories are active research facilities where telescope time is in very high demand (astronomers typically request three to five times as much viewing time as is available), and time on a larger telescope is valued at tens of thousands of dollars per night.

Incidentally, astronomers no longer look through eyepieces on research telescopes, anyway; it's all done with electronic detectors and computers. For the most part, professional astronomers stopped looking through telescopes with their eyes more than 100 years ago, when photographic technology was first developed. Until the advent of electronic detectors, astronomers used photographic emulsions on glass plates to study the universe.

WHAT'S WHAT AND WHERE

The state and federal governments have given a variety of designations to various areas on Mauna Kea. Here's an overview:

The Mauna Kea Forest Reserve (so designated in 1909) is also a National Natural Landmark (designated in 1972). The Forest Reserve, consisting of 83,900 acres, encompasses just about everything above 7,000 feet (2,133 meters).

Between about the 6,000- and 10,000-foot (1,829- and 3,048-meter) levels, a large area of the *māmane-naio* forest is designated as a critical habitat for the endangered *palila* (*Loxioides bailleui*), one of the Hawaiian honeycreeper birds. The endangered state bird, the Hawaiian goose or *nēnē* (*Branta* [*Nesochen*] *sandwicensis*) lives in this area, too.

The Mauna Kea Ice Age Natural

Area Reserve extends from the summit to the south and west down to about 10,500 feet (3,200 meters). It includes such important cultural resources as an ancient Hawaiian adze quarry site, Keanakākoʻi (itself a Natural Historical Landmark), Puʻuhaukea, Puʻu Waiau and Lake Waiau, the only alpine lake in Hawaiʻi. It also harbors rare, native invertebrates and evidence of the ice ages on Mauna Kea.

The Mauna Kea Ice Age Natural Area Reserve is part of the state's Natural Area Reserves System (NARS), a program administered by the Department of Land and Natural Resources, Division of Forestry and Wildlife. The overall NAR System includes 19 reserves on five Hawaiian islands, more than 109,000 acres altogether, and exists to preserve and protect representative samples of Hawaiian biological ecosystems and geological formations.

Above 12,000 feet (3,658 meters), the Mauna Kea Science Reserve—leased from the state since 1968 by the University of Hawaiʻi in order to create a scientific complex for astronomical research—is made up of 11,288 acres surrounding the summit area of Mauna Kea. The Science Reserve is a circular area about five miles (eight kilometers) in diameter where astronomical observatories are clustered. The adze quarry and Lake Waiau are not part of the Science Reserve but are instead part of the Mauna Kea Ice Age Natural Area Reserve.) To the northeast, the Science Reserve extends to include Puʻumākanaka, Puʻu Hoaka and a nearby unnamed cinder cone.

Mauna Kea also has several diverse ecosystem zones. Between 8,528 and 9,512 feet (2,599 and 2,899 meters) it's *māmane* forest. The 9,512- to 10,496-foot (2,899- to 3,199-meter) elevation is alpine scrub, though in the past that may have been the area of Mauna Kea's tree line. Above that, from 10,496 to 11,152 feet (3,199 to 3,399 meters), you'll find scrub desert. The area from 11,152 to 12,464 feet (3,399 to 3,799 meters) is barren alpine cinder desert.

Natural Area Reserves, such as the Mauna Kea Ice Age Natural Area Reserve, are areas of refuge, not recreation, so there are restrictions as to how you can use them. Hiking and nature study are permitted in NAR Systems, as is the hunting of game mammals and game birds in designated hunting areas, subject to the hunting rules of the Department of Land and Natural Resources.

What's not permitted in a Natural Area Reserve, according to the Department of Forest and Wildlife:

Fires, littering, tent camping; injuring, removing or killing any native animal or plant life; vehicles, except on designated roads; removing, damaging, or disturbing any historic or prehistoric remains, or geological feature or substance; introducing any form of plant or animal life; operating a motorized vessel within reserve waters; putting a vessel or material in, on, or otherwise disturbing a lake or pond. ▲

2

The Sacred Mountain

2 The Sacred Mountain

Pāpa-Hānau-Moku, "Pāpa-who-gave-birth-to-the-islands," is the earth mother.

An ancient saying states, *"Mauna Kea kuahiwi kū ha'o i ka mālie"* ("Mauna Kea is the astonishing mountain that stands in the calm"). It suggests the reverence Hawaiians have always had for their tallest mountain.

The earliest known written records, maps, chants and oral histories indicate the mountain has long—perhaps always—been called Mauna Kea. The name Mauna Kea is usually translated as "white mountain," referring to the winter snows that often cover the summit and other high slopes. However, the name Mauna Kea is sometimes also translated as "the mountain of Wākea."

According to some ancient genealogies, Mauna Kea—along with *kalo* (taro), the staple food of Hawai'i, and the Hawaiian people themselves—all descend from the gods Wākea (the expanse of the sky) and Pāpa-Hānau-Moku (the earth mother: literally, "Pāpa-who-gave-birth-to-the-islands"). The mountain has been referred to as Wākea and Pāpa's firstborn child. Hāloa-naka, a stillborn son from whose body the first *kalo* plant grew, and Hāloa, the first person and the ancestor from whom all Hawaiian people descend (sometimes said to have been born of Wākea and his daughter), are also said to have been Wākea and Pāpa's children.

Mauna Kea, then, is actually considered an ancestor of the Hawaiian people: a sacred and respected elder. In the Hawaiian tradition, elders are looked to for wisdom, and this applies to Mauna Kea, as well. There are still those who go up the mountain to pray, seek knowledge and show respect for their ancestors.

In ancient times, before Hawaiians had a system of writing, their genealogies, as well as other histories and stories, were kept by *haku mele* (sometimes translated as "masters of song"). The role of the *haku mele* was to memorize royal genealogies and to compose chants, which were equally carefully memorized and performed. Such a specialist, attached to a royal household, was held in high honor. It was critical that more than one person memorize important genealogical chants, and their careful recitation was extremely important. A breath taken in the wrong place or stumbling over a word was a sign of ill fortune for the person being honored in the chant.

"Hānau A Hua Ka Lani" is one such genealogical chant recounting Mauna Kea as a child born of Wākea. More than 100 lines long and published in a Hawaiian-language newspaper in the early 1900s, it was composed in honor of Kauikeaouli, Kamehameha III:

O hānau ka mauna a Wākea
Born is the mountain sired by Wākea
'Ōpu'u a'e ka mauna a Wākea
Budded forth, the mountain sired by Wākea
'O Wākea ke kāne, 'o Walinu'u ka wahine
Wākea the husband, Walinu'u the wife
Hānau Ho'ohoku, he wahine
Born is Ho'ohoku, a daughter
Hānau Hāloa he ali'i
Born is Hāloa, a chief
Hānau ka mauna
Born was the mountain
He keiki mauna na Wākea
A mountain son sired by Wākea

There's another reason Mauna Kea is considered so important, and more than merely a mountain, to many Hawaiians and in fact to Polynesians as a whole: its summit, the highest point among all the islands of Polynesia, is considered a *wao akua* or sacred realm of the gods. Pu'u Wēkiu (literally, "summit hill") is the name surveyors gave the highest peak. The earlier recorded Hawaiian name for the summit (or perhaps for the entire cluster of cones at the summit) is Kūkahau'ula

Kūkahau'ula is an earlier name of the summit.

This is "the point where the sky and earth pulled apart to create the realms of heaven."

("Kū of the red-hued dew"), also the name of a traditional chief. Some say it is named for the god Kū, a lover of the snow goddess Poliʻahu. It is also said that Mauna Kea's summit is the point where sky and earth pulled apart to create the vast expanse of space and the realms of heaven.

A *wao akua* is usually a remote place where people do not live and, it is believed, spirits do, such as a high mountain or perhaps a desert or deep jungle. These areas, while perhaps not actually *kapu* (forbidden) in ancient times, were generally avoided, probably out of respect.

Mauna Kea is still considered by some Hawaiians to be home to the gods and the ancestors. This is one reason there is controversy over further development of observatories atop the mountain. Building near the peak of what many Hawaiians consider their most sacred place is considered by some to be a desecration of deeply held cultural and spiritual beliefs; others welcome the chance to learn from the exciting and cutting-edge astronomical research being conducted atop the mountain.

Perhaps, for now, the best path available to us is simply to appreciate, respect and *mālama* (take care of) the mountain as best we can.

Nana Veary, author of *Change We Must: My Spiritual Journey*, was a well known Hawaiian *kupuna* (elder) and spiritual leader born at the beginning of the 20th century. She was raised with ancient and traditional Hawaiian values and beliefs, and is quoted in the document "Mauna Kea—the Temple, protecting the sacred resource," giving this simple but wise advice well-suited to visiting Mauna Kea:

> *Ask permission and give thanks—that was the Hawaiian protocol that extended to every aspect of life in nature. If you observe this constantly, you begin to develop an inner silence, a deep strength that comes from having your mind attuned to the universal consciousness that pervades all things.... When I fly to any other island, I ask permission of its guardian spirits. As the airplane lands, I ask permission to be on the island and to partake of its beauty.... I always see a rainbow or some sign of welcome. I always feel this is nature speaking directly to me, responding to my reverence.*

Silently asking permission and giving thanks, as well as exhibiting an overall attitude of respect and having as little impact as possible on the land, are excellent habits to develop—whether on the sacred mountain Mauna Kea, in other places in Hawai'i or elsewhere in the world.

WAHI PANA AND PU'U

Keep in mind that many consider the entire mountain a highly spiritual place, not just the sites noted in this chapter. Many *pu'u* (hills) and other natural features on Mauna Kea not specifically mentioned here are also significant and important cultural sites. Some are *wahi pana* (famed and storied places) described in traditional and historical accounts. An example is those places associated with the snow goddesses.

To understand such accounts of *wahi pana*, you should know that though the word *pu'u* literally means "hill," a *pu'u* is also sometimes considered a *kino lau*. *Kino lau* are forms taken by a god or goddess. A rather well known example is that of the volcano goddess, Pele, who is said to live in the crater Halema'uma'u within Kīlauea's caldera. Pele is sometimes seen in the form of a flame of fire, a young girl or an old woman. Those varied forms are three of her *kino lau*.

On Mauna Kea there are three *pu'u* that are named for, and are *kino lau* of, three sister goddesses—Poli'ahu, the goddess of snow; Līlīnoe, the goddess of mist; and Waiau, goddess of the lake. Each of the goddesses represents a form of water, and all are beauties who wear white robes and are considered wise, fun and full of adventure. All, too, are competitors of Pele.

A fourth sister snow goddess, Kahoupokane, is thought to have been the goddess of the Maui volcano Haleakalā and to control the snows there. Unfortunately, her stories have mostly been lost with time. A spring located on Mauna Kea at 10,500 feet (3,200 meters), Houpo o Kane, is named for her.

Pu'u Poli'ahu is a hill to the west of the summit peak at 13,612 feet (4,149 meters). It's named, of course, for Poli'ahu ("clothed or garment-covered breast"), the goddess of the snows on Mauna Kea and guardian of the mountain. It is Poli'ahu who sometimes "throws her cloak of white" over the summit of Mauna Kea in winter months, creating what we call snow.

Pu'u Līlīnoe, located southeast of the summit peak, rises 12,956 feet (3,949 meters) and is named for the chiefess Līlīnoe, who secluded herself on the mountain and who, after her death, was buried in a cave near the summit. Her importance is signified by the facts that Queen Emma later attempted to collect her remains and that references to Līlīnoe abound in the chants commemorating Queen Emma's visit to the mountain. Līlīnoe is also goddess of the volcano Haleakalā on Maui, as well as of Mauna Kea. Some legends say she was the wife of the great flood survivor Nana-Nu'u, who lived in a cave on a slope of Mauna Kea. She was also Poli'ahu's sister.

Waiau is another snow goddess, though the details of her story have been forgotten. It is almost certainly related to Mauna Kea's Lake Waiau, which is located just above the 6-mile marker on the summit road at the base of Pu'u Waiau. Hawaiians considered Lake Waiau bottomless, and it was believed to extend into the very heart of the mountain.

The goddesses are said to manifest themselves at the *pu'u* named for them. Tread lightly and with respect.

POLI'AHU AND PELE

In her book *Hawaiian Myths of Earth, Sea and Sky*, Vivian L. Thompson recounts a contest on Mauna Kea between Poli'ahu, goddess of snow, and Pele, goddess of fire:

Poli'ahu, the goddess of snow, lived high on the northern slopes of Mauna Kea. One sunny day, she and her friends went down below the snow line with their *hōlua* sleds. Poli'ahu ran, threw herself down on her sled, and flew down the mountain on a thin, grassy *hōlua* track. When she came to a stop, she marked the spot and stepped aside. Her friends followed, flying down the track with their sleds, but no one went as far as Poli'ahu.

At the bottom of the run was a beautiful stranger who stared at Poli'ahu and asked to race her. Poli'ahu welcomed the woman and lent her a sled, and together they climbed the mountain. The stranger ran and then lay down on the sled and flew quickly down the slope. Poli'ahu sledded down the mountain after her, and she went further than did the stranger.

The strange woman was angry and her dark eyes flashed. "The sled was the wrong size for me!" she said.

Poli'ahu found her a longer sled, and again they climbed the hill. Again they raced down the hill on their sleds, and for a second time, Poli'ahu went the greater distance.

The stranger complained again about her sled, so Poli'ahu offered the stranger hers. For their next race, the stranger wanted to go higher up the mountain and race a longer distance. So they climbed as far as the snow line, where they

Poli'ahu "throws her cloak of white" over Mauna Kea, creating what we call snow.

exchanged sleds.

As Poli'ahu started down the mountain, the stranger stomped her foot and the earth shook. Along the lower part of the sled course, a crack opened in the ground. Steam rose from the crack, and suddenly Poli'ahu's friends could no longer see her. When the steam cleared for just a moment they saw Poli'ahu shooting toward the crevice and the stranger just behind her, standing upon her sled. The friends watched as the stranger's black robe turned red and her eyes began to glow.

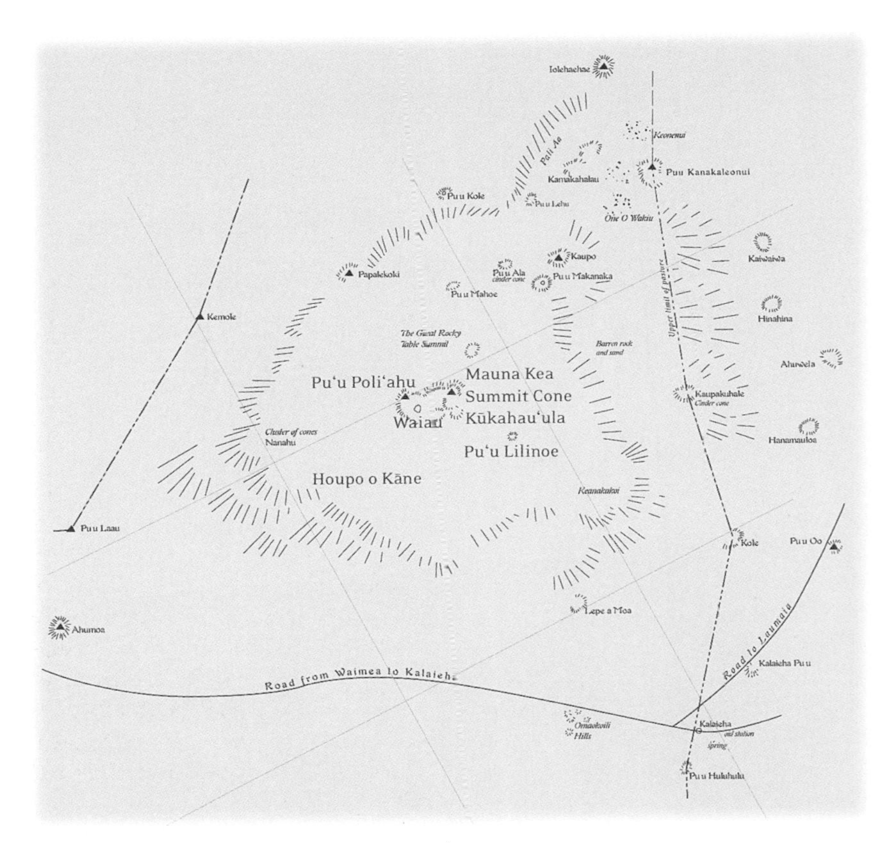

Here are the Hawaiian names of the principle summit features.

Suddenly it was clear: Poli'ahu's competitor was Pele, the goddess of fire!

Pele stomped her foot again and molten lava came shooting up through the crack.

Poli'ahu's friends raised their hands toward Mauna Kea's peak and began to chant. Gray Cloud Goddesses gathered to help Poli'ahu by sending snow from the top of the mountain. The snow fell upon the hot earth and extinguished the lava.

Pele was angry. She cried out and more lava appeared, a row of fountains spouting fire just ahead of Poli'ahu. There was no way to avoid it, and Poli'ahu plunged through the fire. Her golden robe caught fire and she threw it off. Jumping from her sled in her striking white under-robe, she watched calmly as Pele rode toward her atop a red and gold river of lava. Poli'ahu swung her robe and brought an icy wind down from the mountain.

Pele shivered and the fountains of fire died down. The river of lava calmed. Pele screamed for the lava to swallow up Poli'ahu.

But the fountains of fire died, and the lava river slowed further. It flowed to Poli'ahu's feet, where it split in two and went around her. It moved toward the sea where it formed a long flat piece of new land on the eastern side of the island, north of Hilo, still known as Laupāhoehoe, or "Leaf-of-Smooth-Lava," and an arch at Onomea.

Pele was not pleased; she shivered again with cold and disappeared. Pele never again crossed over to Poli'ahu's side of the mountain, though she continued to send her lava down the south side of Mauna Kea.

KEANAKĀKO'I, THE ADZE QUARRY

"O ka po'e ka ko'i, kekahi po'e i mana'o nui ia i ka wā kahiko ma Hawai'i nei" (Adze-makers were an extremely esteemed class in ancient Hawai'i).

David Malo, a Hawaiian scholar born about 1793 and educated at Lahainaluna School on Maui, related this information about the high status of adze-makers. Malo was, and still is, widely respected for his knowledge of ancient Hawai'i and its old ways.

The place where these esteemed adze craftsmen gathered their rock—a material that formed when molten lava erupted under glacial ice caps and which was harder than any other available to them—is on Mauna Kea at about 11,000 to 12,000 feet (3,353 to 3,658 meters). The adze quarry certainly wasn't easily accessible to the workers, nor would it have been comfortable due to the extreme weather at such a high altitude. But people spent time there because the dense lava rock was so valued in making adzes, specialized tools used to cut wood. Adzes were made by chipping a hard piece of rock to a sharp edge, then lashing it to a piece of wood.

The quarry called Keanakāko'i (literally, "the adze-making cave") is on the interior, southwest slope, part of the *ahupua'a* (land division) of Ka'ohe in the Hāmākua district. The site includes quarries, caves where adze-makers slept, adze-making workshops and shrines that workers erected to their gods. Radiocarbon dating from charcoal in the shelter hearths at Keanakāko'i suggests the site was in use from at least the mid-1200s. Keanakāko'i is listed as a National Historic Landmark, is on the State and National Register of Historic Places, and is the largest pre-industrial quarry in the state. Please view it from the road but don't walk to it, as the site is fragile.

Adze-makers, probably from nearby areas, brought their food and firewood with them when they went up the mountain to work the stone for brief periods. For protection from rain they wore capes woven from layers of *ti* leaves. For warmth, they probably also had clothes and bedding made of *kapa* (pounded paper mulberry bark). At the quarry, a man worked the rock by swinging a large hammer stone between his legs and against the edge of a boulder core, a large block of stone or a cliff or outcrop. He examined the flakes that chipped off and selected larger ones as "blanks" for making adzes. He used a smaller hammer stone to "rough out" a blank—remove flakes from both sides—and then carried some down the mountain to a workshop at a lower altitude.

Most stone was worked onsite or at a rock shelter in the main quarry area. Remnants of some of these shelters still exist, as do remains of their shrines, some of their food and offerings, and remnants of the adze manufacturing process itself.

Archaeologists have identified the lower-elevation adze manufacturing workshops by locating byproducts of the adze-making process: flakes, cores, adze rejects and hammer stones. These workshops are different from the ones at Keanakāko'i. There are more adze rejects found at these workshops than flakes, which suggests the adzes were "flaked" (roughly formed) somewhere else and brought to the workshops for finer work. Some of these workshops have one or more shrines, and some shrines have adze rejects and flakes that seem to have been placed deliberately. Hawaiian scholar David Malo interpreted these as offerings to the gods of adze-making.

At the workshop, a craftsman sharpened the adze against a wet slab of fine-grained stone. An hour or two of grinding against the stone produced an edge with a sharp, tapered blade. The stone was placed against a cushion of *kapa* (bark cloth) to absorb shock and then lashed with braided sennit to a carefully selected haft. Often cut from a *māmane* tree, the chosen piece of wood usually had two sections with a 70-degree angle between them.

Archaeologists say that Keanakāko'i produced far more adzes than could have been used locally. Perhaps they were traded over a wide distance within the Hawaiian Islands or maybe even on voyages to Tahiti or elsewhere in Eastern Polynesia.

The adze quarry seems to have been out of use by 1800, probably due to the introduction of iron to Hawai'i.

The area is designated by the state as a Natural Area Reserve, and the removal of any material is a violation of state law.

FEATHERS

Another Mauna Kea resource important to early Hawaiians was the colorful bird feathers used to make ceremonial capes, *kāhili* and helmets for Hawaiian royalty. Some capes contained tens of thousands of feathers. It's thought that certain birds in wetter forests on the lower slopes of Mauna Kea were hunted for their feathers, while on the upper slopes of the mountain, adze-makers and others frequently caught the *'ua'u* (dark-rumped petrel) for food; its bones are common in the midden deposits of rock shelters in the quarry.

ARCHAEOLOGY

In the 1980s and '90s, archae-

Many shrines and other religious sites are subtle and can be inadvertently destroyed.

ologists identified 93 sites of note in the Mauna Kea Science Reserve. That research project surveyed only about a quarter of the total reserve; there are almost certainly many other sites still to be discovered.

It's a delicate subject—one person's archaeological find can literally be another's esteemed ancestors. Many sites on Mauna Kea tell stories of work and religious observance on the mountain that predate first Western contact in 1778. Some are actual burial areas, containing bones of Hawaiian people laid to rest on Mauna Kea. These sites are, obviously, of great cultural, spiritual and familial significance.

How do you respect this balance when you visit Mauna Kea? Start by staying on the road or trail at all times. Many shrines and other ancient and contemporary religious sites, which are made of stone, are subtle and not readily noticed by the casual visitor, and can easily be destroyed by hikers inadvertently tromping over them.

SHRINES

Many of the archaeological sites identified on the mountain to date are said to be *kuahu* or shrines, which were places of worship. Sir Peter Buck, an anthropologist and a former director of Honolulu's Bishop Museum, defined *kuahu* as "a simple altar without a prepared court. They were made by individuals or small family groups who conducted a short ritual that required no priest."

Each upright stone in a shrine may symbolize a different god.

Because there are neither organic deposits nor any other material datable through modern methods present at these shrines, we cannot discover how old they are. Archaeologists speculate that some may date from 700 to 800 AD, though they say most probably date to around 1400 to 1600 AD. That's when the population of the island increased, and it's also when there was greater activity in terms of adze production at Keanakāko'i.

Shrines on Mauna Kea always contain one or more upright stones, though they are arranged in different patterns and rest on different types of foundations. Some sit on bedrock, some on boulders, and others on platforms, mounds or cairns.

They vary in design, as well. Some are simple while others have "pavements" of stones that create a level surface. These "pavements" are thought to have been where people placed offerings; a priest took the offering from there and ceremonially presented it to the gods. Some shrines have "prepared courts" and might be called *heiau* (temples), though there is some disagreement about this. "Courts" were sites with a pavement on one side of a row of altar uprights.

One of the first Westerners to describe the upright stones at shrines was Kenneth Emory, a Bishop Museum anthropologist who studied the adze quarry area in 1937. Emory theorized that each upright stone symbolized a different god. He pointed out the similarities between the way the Hawaiians and other island groups who share a common Eastern Polynesian background honored their gods—the latter have similar structures, also with upright stones, that are known to be shrines:

The adze makers, clinging to the ancient form of shrine at which to approach their patron gods, have preserved a most important link with their ancestral home. Each upright stone at a shrine probably stood for a separate god. The Hawaiian dictionary describes 'eho as "a collection of stone gods" and this is the term which the Tuamotuans, the neighbors of the Tahitians, used to designate the alignment of upright stones on the low and narrow platform at their maraes, or sacred places.

A rock cairn may denote a place of worship.

Sir Peter Buck agreed that the upright stones were probably representations of gods. He pointed out that people on other Polynesian islands used stones to represent gods, be they gods of a craft (like adze-making) or *'aumakua* (family or personal gods). Some people used unfinished stone, while others made rough human forms from their stone. It was not, though, the shape or finishing of the rock that gave it its power, according to Buck—it was the prayers and offerings that gave the stone its *mana* (spiritual strength).

In Hawai'i, the shape of an upright stone is thought to indicate not only a particular god but possibly also whether it was male or female. An unworked stone that stands high and straight, for example, represents the god Kū's virility.

Slab-shaped or upright stones are called *pōhaku o Kāne,* representing the god Kāne, and are male. Flat or rounded ones, female, are *pāpa-o-Hina* or *pōhaku o Hina* and represent the goddess Hina. In addition, the gods Kāne and Hina are thought to have stone children. Some say that a pebble found between a male and a female stone will get bigger and bigger until it is taken to a *heiau* to be made a god in its own right.

Ancient offerings have not been found on shrines at Mauna Kea, aside from adze material at a few and a single *'opihi* (limpet) shell at one site that must, of course, have been carried by someone up from the ocean. It's possible rituals were performed without tangible offerings being made, or that offerings were of perishable materials, not ones such as coral, shell, bone or stone that would last to this day.

BURIALS

Pre-contact Hawaiians took great care to hide the bones of their dead so they could not be defiled. It was not uncommon in the past for Hawaiians to bury their dead in high, remote and sacred places such as Mauna Kea. Oral and family histories contain many references to people being buried on the mountain.

Several sites on Mauna Kea have been identified in recent times as possible burial places,

This shrine was probably erected by the adze-makers who worked in the quarry.

Shrines may be found on platforms, mounds or cairns.

though human remains have not always been discovered there. The wind has uncovered some shallow graves on Mauna Kea, and there are thought to be many more unmarked ones.

Some Hawaiian families continue this tradition of burial on Mauna Kea by taking cremated remains up the mountain.

THE SUMMIT

Aside from one cairn spotted by Reverend Joseph Goodrich during an 1823 climb, no lasting evidence of human activity from pre-Western contact times has been found at Kūkahau'ula, the summit. People have long speculated about this lack of evidence. In the early nineteenth century, some Westerners assumed Hawaiians avoided the summit of Mauna Kea both because of the extreme cold and due to superstitious beliefs.

The missionary William Ellis, in his *Journal of William Ellis*, wrote:

The snow on the summit of the mountain, in all probability, induced the natives to call it Mouna-Kea (mountain white), or, as we should say, white mountain. They have numerous fabulous tales relative to its being the abode of the gods, and none ever approach the summit—as, they say, some who have gone there have been turned to stone. We do not know that any have been frozen to death; but neither Mr. Goodrich, nor Dr. Blatchely and his companion,

Mauna Kea remains a site of traditional rituals and worship today.

could persuade the natives, whom they engaged as guides up the side of the mountain, to go near its summit.

It's unlikely that Hawaiians stayed away from the summit solely because of cold weather, as there are numerous shrines located just a few hundred feet below the peak. It's much more likely that Hawaiians stayed away from the summit, the home of their gods, out of respect. Possibly it was under a *kapu*, and thus restricted and accessible only to the highest chiefs or priests.

EARLY EXPLORERS

While Hawaiians, of course, climbed Mauna Kea, long before the first Westerner, most of their stories were never written down, and many have been forgotten. Some of the first *malihini* (newcomers) to climb Hawai'i's mountains wrote about their expeditions in journals.

1794

Just 16 years after Westerners first arrived in Hawai'i, the botanist Archibald Menzies, who arrived with the English explorer Captain Vancouver, climbed not Mauna Kea but Mauna Loa. This was the first documented climb of one of the Big Island's mountains by a foreigner.

1823

The missionary Joseph Goodrich, who reached the summit in August and experienced altitude sickness along the way, seems to have been the first Westerner to climb Mauna Kea, and he wrote in his journal about his expedition.

Imagine Goodrich's laborious journey. Instead of cruising up the mountain in a car with air-conditioning and heating, as we do now, he and others of his era hiked laboriously on foot or rode a horse or mule. They didn't follow carefully maintained paved roads, as we do, but forged their way across loose cinder while hauling their own provisions, including food and water, up the steep slopes.

The English missionary William Ellis wrote this about Goodrich's climb:

At noon he dismissed his native companion, and, taking his great coat and blanket, began to ascend the more steep and rugged parts. The way

Trekking up volcanoes has always been an exciting activity. Here early visitors to Kīlauea prove they were there by scorching postcards to send home.

This is believed to be the 1925-27 Geological Survey Team.

was difficult, on account of the rugged volcanic rocks and stunted shrubs that covered the sides of the mountain. On his way, he found numbers of red and white raspberry bushes loaded with delicious fruit.

At the summit, Goodrich found a rock cairn left by an earlier visitor. It's not known who erected the cairn, or when.

1825

James Macrae climbed Mauna Kea and stayed for two days. The botanist from the *H.M.S. Blonde* wrote of the lava rocks he encountered there, "[they] have beyond a doubt been thrown up by some previous convulsion."

1830

Accompanied by the American missionary Hiram Bingham, Kauikeaouli, Kamehameha III, visited the mountain on horseback.

1834

The botanist David Douglas climbed the mountain in January, noting the summit's cinder cones and its desolate landscape. He mentioned "a death-like stillness...not an animal or insect to be seen...not a blade of grass." He also wrote that the alpine level of the mountain was carpeted with wild strawberry plants.

1841

Dr. Charles Pickering and William D. Brackenridge of the U.S. Exploring Expedition climbed Mauna Kea.

1853

George Washington Bates hiked to Mauna Kea's summit and wrote that along the way he saw huge strawberry beds and tall raspberry bushes that bore incredibly big fruit. Surrounding the berries and eagerly feeding were native birds, he wrote, including many *nēnē*. At night, he made a fire with piles of withered silverswords.

1873

Isabella Bird, a well-traveled visitor to Hawai'i who later wrote *Six Months in the Sandwich Islands*, visited Mauna Kea.

1882

After spending a day at the summit, C.A. Dutton, the first geologist to study the summit area, described the basalt blocks he found there as having been shaped by frost and ice.

1883

Queen Emma traveled over Mauna Kea en route to Waimea; nine years later W.D. Alexander saw a pillar or cairn that had been built to commemorate her visit.

1892

W.D. Alexander led a surveying expedition up the mountain. They spent time on the summit of Pu'u Līlīnoe, noting stone cairns and interesting land formations. The same year, in a *Hawaiian Gazette* article entitled "The Ascent of Mauna Kea, Hawaii," Alexander wrote:

On Monday, the 25th, the thermometer stood at 20 deg. at sunrise. Messrs. Muir and Alexander ascended the second highest peak on the northwest, overlooking Waimea, 13645 feet height to continue their survey. In the cairn on the summit a tin can was found that contains brief records of the visits of five different parties from 1870 to the present time, to which we added our own. A party of eight girls from Hilo, "personally conducted" by Dr. Wetmore and D.H. Hitchcock, Esq, in 1876, must have been a merry one. Cpt. Long of H.B.M.'s Ship Fantome had visited this spot in 1876, and Dr. Arning with several Kohala residents in 1885.

1909

Though Dutton had pointed out the influence of ice on rocks at the summit when he visited in 1882, it was Harvard geologist R.A. Daly who first seems to have recognized and explicitly pointed out that glaciation had occurred on Mauna Kea: "Hawaii itself seems to have borne at least one small glacier, the characteristic traces of which were observed by the writer on Mauna Kea at the 12,000 foot level."

1925-27

The U.S. Geological Survey first mapped the mountain. ▲

NATURAL HISTORY 3

3 Natural History

GEOLOGY

The Hawaiian archipelago consists of more than 50 huge volcanoes, most entirely beneath the sea. Together they extend 2,200 miles (3,540 kilometers) along the Pacific Ocean floor. Some are as large as 75 miles (120 kilometers) in diameter and stand 30,000 feet (9,144 meters) tall. Volcanoes that extend above sea level make up the eight main islands of Hawai'i.

One such volcano is Mauna Kea, which measures about 15,000 feet (4,572 meters) from the ocean floor to sea level, and then almost the same height again—13,796 more feet (4,205 meters)—from sea level to its summit. Measured from its base, Mauna Kea is the tallest mountain in the world. It is 51 miles (82 kilometers) long and 25 miles (40 kilometers) wide. On the ground, the volcano and its lava encompass about 920 square miles (1,480 square kilometers), comprising almost a quarter of the Big Island.

Along with Mauna Kea, four other mountains make up the landmass that is the island of Hawai'i. Kohala (5,480 feet or 1,670 meters high) is the oldest. Neither it nor Mauna Kea has erupted since writ-

The Big Island is still growing today.

ten records have been kept—Mauna Kea's last eruption was about 4,500 years ago, and it is classified not extinct but "long dormant." Hualālai (8,271 feet or 2,521 meters) most recently erupted in 1801. Mauna Loa (13,677 feet or 4,169 meters), the volcano across the "saddle" from Mauna Kea, is one of the most active volcanoes in the world. Its number of eruptions, though, has decreased in recent decades and its last eruption was in March and April of 1984. Kīlauea (4,090 feet or 1,247 meters) is a young volcano that has been almost continuously active since people first arrived on this island.

And the process continues: a new volcano is forming about 18 miles (29 kilometers) to the southeast of the island. Named Lō'ihi ("Long") by scientists, the volcano, presently 15 miles (24 kilometers) long and eight miles (13 kilometers) wide, is building up from the ocean floor through its eruptions. It presumably sits above the "hot spot" thought to have produced the island chain.

This newest volcano rises some 12,000 feet (3,658 meters) above the ocean floor with about 3,000 feet (914 meters)—and probably thousands of years—to go before it breaks through the ocean's surface. On Lō'ihi's summit is a caldera with two craters. Flurries of earthquake activity have alerted scientists to possible eruptions, and research vessels have gone down to photograph and study the underwater volcano.

A group of highly respected Big Island Hawaiian cultural practitioners call the new volcano not Lō'ihi but Kama'ehu ("The Red Child"), which they say is more appropriate for an island being born under the sea through volcanic activity. They have petitioned to have the name changed officially. The proposed name comes from an ancient chant that speaks of the "reddish child of Kanaloa," Kanaloa being the god of the ocean.

Under the sea, another island is being born.

Mauna Kea and its lava flows extend all the way to the sea beyond Route 19, which runs along the Hāmākua Coast. Heading north from Hilo, you drive across the flanks of Mauna Kea for 40 miles (64 kilometers) until reaching Honoka'a. To the northwest, lava flows from Mauna Kea meet those of Kohala Volcano at about 3,000 feet

(914 meters). As you drive along Saddle Road between Mauna Kea and Mauna Loa, you'll see more recent lava flows, from Mauna Loa, that as recently as 1936 covered older Mauna Kea lava.

HOW HAWAIIAN VOLCANOES FORM

In 1910, an Austrian meteorologist named Alfred Wegener theorized that the continents had once been joined and had broken apart. He pointed out that North America's eastern coast fit almost like a jigsaw puzzle piece to Europe's western one, as did the coast of South America to that of Africa. He also noted how the old mountain ranges of eastern North America, such as the Appalachians, seem to "continue" in very similar ranges across the ocean in Ireland, Scotland and Scandinavia. At the time, however, tests proved that the ocean floor was too hard to permit the continents to move through it, and his theory of "continental drift" was discredited.

In the 1960s, though, research by oceanographers revealed that new ocean floor is created alongside underwater mountain ranges. Mountain chains seem to form where the oceanic plates to which continents are attached meet. The oceanic plates drift about an inch (a few centimeters) per year, and the cracks where they split apart are filled in with molten lava from below. Wegener's theory was revisited, with a new idea that the continents are attached to plates of the spreading sea floor that move with—instead of through—the ocean crust.

In places where Hawaiian volcanoes form, the Pacific plate sits atop an ocean floor layer that is itself over a "hot spot," or place of abnormally hot rock. While the ocean floor drifts almost imperceptibly to the northwest, the hot spot is fixed in place. Molten rock spews out from this hot spot, causing a volcano to form underwater. Lava continues to build up until finally the volcano is high enough to emerge from the sea. This building-up process continues until the drift of the ocean plate moves beyond the hot spot, at which time that volcano stops receiving new magma from the hot spot. Eventually a new one begins forming.

Scientists studying the age of volcanoes have estimated that since the island of Kaua'i was formed, the Pacific plate on which the Hawaiian Islands sit has moved to the northwest approximately four inches (10 centimeters) per year—about the rate at which a human fingernail grows.

This movement explains why the volcanoes at the northwest end of the Hawaiian chain are the oldest ones and the ones at the southeast end, such as the Big Island, which still sits mostly above the hot spot, and emerging Lō'ihi, are the newest.

Interestingly, this is what Hawaiian lore has said all along—that Pele, the goddess of fire whose lava creates new land, started out on Kaua'i, then made her way to O'ahu

The first growth emerges after a lava flow.

and Maui, ending up (and presently living) in Halema'uma'u Crater at the very active Kīlauea Volcano on the Big Island. This traditional story of Pele's journeys through the island chain correlates with the ages of the islands.

The other volcanoes on the main Hawaiian Islands are extinct, except Haleakalā, though they're still young enough that they remain high-standing islands. Volcanoes that are older still, such as in the northwestern Hawaiian Islands, have eroded and sunk; some now form atolls and reefs. The oldest are completely submerged beneath the sea.

Dating volcanic rock by radioactive potassium-argon techniques reveals that the oldest Hawaiian volcanoes on the main islands, those on Kaua'i, are about 5.6 million years old. Mauna Kea is about one million years old. Both are young in geologic terms: the Rocky Mountains and Sierra Nevada mountains are both millions of years older.

ROCKS

Lava starts out below the earth's surface as magma. When it spews out to form a volcano it goes through changes in its chemical makeup. The first type of lava emitted is called *tholeiitic basalt* and is a heavy, dark rock rich in iron and poor in silicates. That fine-grained lava rock, so common in Hawai'i, often contains small green olivine crystals. As the volcano gets bigger, the composition of the lava changes, and it produces a lighter-colored, usually brownish rock called *andesite*. In Hawai'i this is known more specifically as *hawaiite* or *mugearite*. It's slightly richer in silica than basalt and has more alkali minerals.

Geologists divide the lava rock of Mauna Kea into two types, based on chemistry and age. The bulk of the mountain is older rock, perhaps 100,000 to 150,000 years old, primarily made up of primitive olivine basalt they call the "Hāmākua formation." Eruptions of newer lava, called the "Laupāhoehoe series," began on Mauna Kea about 65,000 years ago. These are mostly andesine andesites ("hawaiites"), and form a thin layer over the upper part of the mountain. Most of Mauna Kea's lava flows are made up of *'a'ā* lava, which has a rough surface as opposed to *pāhoehoe's* smooth, ropy one.

Unlike the younger Mauna Loa, with its smooth long profile and scarcity of cinder cones at the summit, Mauna Kea has almost 100 cinder cones on its upper slopes, many asymmetrical because of the winds that blew most of the ash and cinder to the southwest. The cones lie in a pattern that indicates the volcano formed over rifts that extend eastward, southward and westward. The cinder cones are another clue to Mauna Kea's older age—they were created by the eruptions that happen not on a younger volcano but on a mature one.

The most recent eruptions of Mauna Kea account for the small cinder cones you'll see south of the observatories.

Cinder cones are made up of ash and cinder as well as volcanic "bombs," tephra and splattered bits of lava. Volcanic bombs are elongated, almond-shaped bits that blast out of a volcanic vent and solidify in the air. "Tephra" is a general term for material blasted into the air by an eruption.

SHIELD VOLCANOES

The magma under Hawai'i's volcanoes is some of the hottest on earth (at Kīlauea Volcano, it's been measured at 2,400 degrees Fahrenheit (1,316 degrees Celsius). Because of its high temperature, chemical makeup and gas content, it pours out as a fluid, honey-thick lava that can flow great distances.

Hawai'i's shield volcanoes have long, smooth ascensions.

(It's called "magma" when it's underground and "lava" above the earth's surface.) Layers of these thin lava flows build atop one another, creating rounded, shield-shaped mountains. If the resulting mound measures just a few miles across, it's called a "lava shield." When it's tens of miles in diameter, it's a "shield volcano." The names come from an Icelandic volcano said to resemble a warrior's round shield lying face-up on the ground; in the 1800s, a German geologist visiting Iceland picked up on the name and applied it to all similarly shaped volcanoes. There is not a lot of explosive activity on a shield volcano.

All Hawaiian volcanoes start as shield volcanoes. Mauna Loa and Kīlauea are good examples of how a shield volcano looks. Their long, smooth ascensions—as opposed to the more typical, conical look of, say, Mt. Fuji in Japan—often perplex visitors expecting a more stereotypically shaped volcano.

Mauna Kea is an older volcano—it's estimated to be about a million years old—and more than 200,000 or 250,000 years ago it moved from being a shield volcano into a "post-shield" stage. The other older volcanoes on the island, Huālalai and Kohala, are also post-shield ones. Older, post-shield volcanoes have very low rates of eruption compared to younger volcanoes. Their lava has a different chemical makeup and the volcanoes themselves are steeper, with less regular topography. In addition, post-shield volcanoes do not have calderas at their summits.

CALDERAS

Normally a Hawaiian volcano has a large caldera (a pothole-shaped depression roughly circular in shape with a flat floor, bounded by steep cliffs) at its summit. Calderas are formed when the surface of a very large crater collapses into an emptying magma chamber. On this island, both Mauna Loa and Kīlauea have large calderas.

You won't see a caldera on Mauna Kea, although geologists have found evidence of one in the basaltic shield that makes up the mountain's core. It's been almost completely obliterated and buried now by more recent andesite rocks. According to *Mauna Kea: A Guide to the Upper Slopes and Observatories* (Dale P. Cruikshank), in some places, such as Pōhakuloa Gulch, where the basalt shield is exposed, it shows evidence that the caldera may have been about 700 feet (213 meters) deep. Its rim is at about the 12,700-foot (3,871-meter) elevation. The caldera was filled in by later andesitic lava and by the action of ice during Mauna Kea's glacial period.

PIT CRATERS

A pit crater is similar to, but smaller than, a caldera and is also formed by collapse. The one pit crater in Mauna Kea's summit area is located within the Mauna Kea Ice Age Natural Area Reserve. There's a good view of it from the slope above, near the adze quarry and east of Pu'uko'oko'olau.

SUMMIT ROAD GEOLOGY

It's about 15 miles (24 kilometers) from the Mauna Kea turnoff at Saddle Road to the summit area of the mountain. About 5.5 miles (8.5 kilometers) up from Saddle Road, you'll see two cinder cones to the east. Pu'ukole, which erupted about 4,400 years ago, is the younger of

Mauna Kea is a volcano born in fire and now often capped with snow.

the two. In the foreground is an older flow, about 5,300 years old, which spewed from a mound about a half-mile up the slope. You'll also pass the horseshoe-shaped crater called Pu'ukalepeamoa, which is just below Hale Pōhaku (at the Visitor Information Station level) on the opposite side of the road. The ridge on the road's west side is the crater's eastern rim, where cinder and fragments of older rock can be seen.

Hale Pōhaku, six miles (9.5 kilometers) up and at the 9,200-foot (2,804-meter) level, sits on a smooth layer of ash and cinder. Between Hale Pōhaku and the summit you'll see *'a'ā* lava on the south and southwest sides of the mountain. As you drive beyond Hale Pōhaku, you climb a steep cinder slope via five switchbacks that ascend about 2,000 feet (610 meters). At the top of the switchbacks, you're at the eastern base of a cinder cone called Pu'u Keonehehe'e. To the southeast is one of the few pit craters on Mauna Kea. Above that, the road takes you into what was once glacier territory.

At 12.1 miles, you'll see Pu'u Wēkiu, the volcano's summit cone, and the observatory buildings perched on its adjacent ridge. Watch, for about a mile after this, as the road follows a flow that came from that *pu'u*. In places you'll see grooves in the rock. These striations were formed by a glacier flowing over the ground; the rocks imbedded in the bottom of the glacier cut these grooves.

At 12.5 miles, you'll encounter a small parking area. Across the road is a trail leading west toward Lake Waiau, about a half-mile away. The path runs between the cinder cones Pu'u Waiau to the south and Pu'uhaukea to the north. Lake Waiau sits at the bottom of Pu'u Waiau's crater—it's approximately a 30-minute hike each way.

The summit road ends at the rim of Pu'u Wēkiu. The summit sits on the eastern rim of Pu'u Wēkiu's crater, about 600 feet (183 meters) southeast of the parking area. The observatories are immediately before you, and the view out from the mountain is striking: look south from the parking area and you can see Moku'āweoweo Caldera on Mauna Loa. To the southwest is Hualālai. Kohala is to the north. And looking to the northwest, you can see the volcano Haleakalā on Maui, and occasionally the West Maui Mountains and the islands of Moloka'i and Lana'i.

Algae gives Lake Waiau a vivid green color.

LAKE WAIAU

The nearly round alpine lake at the base of Pu'u Waiau crater is about a 30-minute walk from the summit road at mile 12.5 (as measured from the Saddle Road turnoff). The lake is rich with algae, which gives the water a vivid green color.

The permanent, 1.5-acre lake southwest of the summit is the highest lake in the Pacific Ocean basin, and at 13,020 feet (3,968 meters) it's one of the highest in the United States. The lake's water comes from melted snow and precipitation. If the lake bottom were lava rock, as one might suspect, it wouldn't hold water. The reason it does is that the lake floor is made of clay formed from ash that erupted from Mauna Kea 3,300 years ago.

Lake Waiau, thought by many early Hawaiians to be bottomless, is a *wahi pana*, a sacred place accorded

Practitioners of lā'au lapa'au *collect water from the lake for traditional healing rituals.*

great respect. Its waters are also known as the "sacred waters of the god Kāne"; the word *waiau* means "swirling waters." Lake Waiau is considered to be the *piko* (umbilical cord) that connects the islands to the heavens. Some Hawaiians used to ascend the mountain to offer the *piko* of their newborn children into the lake as a way of safeguarding the infant's future, and some continue to do so today. Practitioners of *lā'au lapa'au* (native healing) have always collected water from the lake for use in traditional healing techniques. Other contemporary practitioners do as well, and the water is used for medicine, blessings and cleansings.

Hawaiians say that water from Mauna Kea runs through the ancient *'auwai* systems (waterways for irrigating taro) that still exist in Hilo today.

HAWAI'I'S ICE AGE

It may be difficult to believe, but Mauna Kea used to have glaciers at its summit. The most recent ice cap on the mountain, which covered about 28 square miles (45 square kilometers) of the summit, was probably at its thickest (about 200 to 350 feet or 61 to 107 meters thick) about 20,000 years ago, and had melted by 10,000 years ago. Several cinder cones are tall enough to have peeked out above the ice cap. That most recent ice cap covered Mauna Kea at a time when Northern Europe and much of Northern Asia were also covered by ice, as well as almost everything above the latitude of New York City. Evidence suggests that there have been two full-fledged glacial periods on Mauna Kea: the earlier between 70,000 and 150,000 years ago, and the later between 13,000 and 40,000 years ago.

Extending as far down as Pōhakuloa Gulch at 10,500 feet (3,200 meters), the glaciers scraped some areas clean of ash and cinder. At the edges of the ice cap area are moraines—mounds of debris picked up by glaciers as they moved across lava flows and cinder cones. These can be seen from the Saddle Road. There is an especially visible area, a V-shaped moraine left by a glacier, which can be seen at the head of the gorge cut by Pōhakuloa Stream.

You can see small areas of striation on rock ledges at the summit, where lava pieces carried by a moving glacier polished the rock beneath the ice. Other rocks, polished into rounded knobs, are called by the French term *"roche moutonnée"* ("fleecy rock") because they resemble the rounded forms of sheep.

There is also evidence of glacial activity in rock outcrops and sides of *pu'u* in the summit area, and there are large areas of glacial till (a fine material mixed with blocks that was deposited by moving glaciers).

Two cinder cones at the summit, where temperatures drop below freezing, show evidence of permafrost in that they have ground ice buried within their craters. The permafrost layer, 35 feet (10.5 meters) thick in places, sits about 16 inches (40.5 centimeters) below the surface. Cycles of freezing and thawing continue, and these cause changes to the patterns of rock fragments.

Even during Mauna Kea's glacial period, the volcano was erupting, sending lava flowing beneath the ice cap. New cinder cones might have been formed when lava flowed beneath the ice, melting the ice and eventually poking through. The edges of these flows cooled quickly when they hit melted ice and formed a uniquely hard and dense rock. This is the rock that, thousands of years later, Hawaiians used to craft adzes.

Most of the glacial features on the mountain, as well as the adze quarry, Lake Waiau, and Pōhakuloa Gulch, are in the area designated as the Mauna Kea Ice Age Natural Area Reserve. When you're driving to the summit, you'll see this as a large area to the left of the road between 10,400 and 13,200 feet (3,170 and 4,023 meters) (about 1.5 through 7 miles up from the Visitor Information Station; signs along the road mark the area). This is a conservation area and is highly protected.

Geologist Gordon MacDonald describes how glaciers could again form in Hawai'i: he estimates that if Hawai'i had just two more inches (five centimeters) of rainfall per year, or if average temperatures dropped just a few degrees, the mountain's high peaks would be covered with snow year round, and a glacier would soon form. Glaciers occur when yearly snowfall doesn't have a chance to melt before more falls. As the layers of snow get deeper and heavier, the lower layers are eventually compacted into a dense blue ice. The weight of the ice causes it to flow slowly. It wouldn't be completely far-fetched to find glaciers in Hawai'i—there are glaciers today in the Andes, which is an equatorial region (though they're south of the

Cracks and crevices shelter plants at the summit from cold and wind.

There are two lichen species found only on Mauna Kea.

tropics), and in the Himalayas, which lie in the tropical belt of Asia.

FLORA AND FAUNA

With its unusual history of both volcanic and glacial activity, Mauna Kea offers a unique habitat for plants and animals. Near sea level, some of the mountain's lava flows lie in dense rain forest. As you follow the lava up the mountain you pass scrub forests, barren lava flows and grassy pastures. Up top, a mile above the clouds, the alpine desert looks as though it could be a moonscape.

On some of the lower flanks of the mountain, to about 6,000 feet (1,829 meters), there are *koa (Acacia koa)* and native *'ōhi'a (Metrosideros polymorpha)* trees, along with ferns and shrubs that grow beneath them. From 6,000 to about 10,000 feet (1,829 to 3,048 meters), the *'ōhi'a* disappears and the terrain becomes open woodland with *māmane* (*Sophora chrysophylla*) and *naio* (*Myoporum sandwicense*), both endemic to Hawai'i, as well as *koa*. There are also two species of mint and native shrubs and vines.

At about the 10,000-foot (3,048-meter) level, the open woodland ends and you find bare lava and cinder slopes with occasional *pūkiawe (Styphelia tameiamiae)* and *'ōhelo (Vaccinium reticulatum)* shrubs. Above 11,500 feet (3,505 meters) there are mostly lichens, mosses and ferns that grow protected in cracks and crevices. Though they aren't obvious unless you're looking for them, about 12 species of moss grow on Mauna Kea, along with 25 different lichens, six species of vascular plants and a red snow alga. About half the lichen species are endemic to Hawai'i, and two of them occur only on Mauna Kea.

SILVERSWORD

Mauna Kea silverswords *(Argyroxiphium sandwicense* ssp.*sandwicense)*, a subspecies found only on Mauna Kea's upper elevations at around 8,500 to 12,500 feet (2,591 to 3,810 meters), are remarkable in that each plant takes up to 50 years to flower—which it does just once. Then, having produced seeds for the next generation, it dies.

Silverswords, or *'āhinahina* in Hawaiian (literally, "silver"; also, Hina, mother of the demigod Maui, is a Polynesian goddess associated with the moon), are beautiful plants with lovely long, sharp and shimmering silver leaves. Each plant sends up, as its crowning achievement, just one five- to six-foot stalk with hundreds of one-inch blossoms that look like tiny sunflowers.

You can see 25 to 50 silverswords that have been planted near the Visitor Information Station in the Hale Pōhaku enclosure. The plants with the green flags were planted in 2003, and the yellow flags indicate plants established in 2004. It takes perhaps 10 to 20 minutes to walk through the enclosure, the entrance gate of which is about 40 feet from the back of the VIS parking lot. You'll also see a *kuahu lele* (Hawaiian altar) where ceremonies are occasionally held. Please don't touch the flowers or the altar, and don't walk or stand too close to the plants, as their roots are susceptible to damage from foot traffic.

Many silverswords are pollinated by hand to ensure genetic diversity.

It's thought that in pre-Western-contact times there were tens of thousands of these unique, showy plants growing on the mountain, including at the summit area. Hawaiians sometimes used the

Scientists fence off silverswords to protect them from grazing animals.

pounded leaves of the silversword to treat asthma.

The plant's decline started in the late 1700s, when goats and sheep were introduced to the island. The animals found the silverswords tasty, and as the herds' numbers increased to tens of thousands, the silverswords' numbers decreased almost to the point of extinction.

Things started turning around in the 1970s. One person gathered seeds from two silverswords on the mountain that were then germinated and grown in a nursery. The state's Division of Forestry and Wildlife fenced in some of the last remaining silverswords on Mauna Kea in order to protect them from grazing animals. The state began planting silversword seedlings in

Efforts to bring the silversword back to its former glory have been successful thus far.

the fenced reserve.

Despite these efforts, in 1986 the silversword was listed by the federal government as an endangered species.

In the 1990s, there were renewed efforts to bring back the plant. More were planted on the mountain, but these unexpectedly developed multiple flowering heads, which is not the norm. Botanists speculate that this was due either to the seeds' having been grown in a greenhouse or to the effect of all the propagated plants' coming from only one or two female parent plants. This lack of genetic diversity did not bode well for the success of the silverswords, so scientists went looking for naturally occurring silverswords. They found fewer than 50.

In order to flower, a silversword must be pollinated with pollen from another silversword. This was a problem because there were so few silverswords, and they weren't close enough to pollinate each other naturally. Then, too, their flowering schedules were infrequent and not necessarily synchronous. So scientists stepped in. Now they watch for blooms, then collect pollen with wax paper and carefully funnel it into vials. Then, using a paintbrush, they painstakingly "paint" the pollen onto other silverswords. Seeds from these plants are germinated for later replanting.

It's a tricky business: the few silverswords that remain have survived because they grow in places where sheep and goats can't get to them. This means scientists find themselves rappelling down dangerously steep cliffs and into crevices. After planting the seeds, they often hand-water them, hauling water up the mountain to high altitudes where just walking takes extra energy.

Today there are again thousands of silverswords on Mauna Kea.

All their efforts seem to be working, though. More than 90 percent of the transplanted seedlings survive now, and today there are again thousands of silverswords growing on Mauna Kea.

The silversword's continued success on the mountain remains to be seen. It is naturally pollinated by the yellow-faced bee, which is endemic to Hawai'i and is declining in numbers due to the non-native ants that eat its larvae. With fewer bees to pollinate the silverswords scientists can't get to, there is no guarantee that silverswords will continue to reproduce in such numbers.

WĒKIU

Most of the creatures living at the summit, including four species of spiders and five species of mites, are arthropods. There are also centipedes, moths, springtails, bugs, booklice, barklice, beetles, wasps, butterflies and flies living in the high-altitude cinder and soil. Some are probably blown to the summit by strong winds, while others are thought to be endemic to the area.

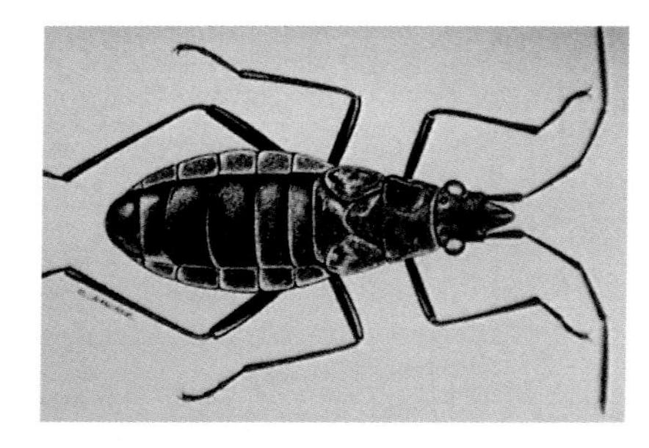

It's thought the wēkiu *has an antifreeze-like compound in its blood to help it survive the cold.*

The *wēkiu* (*Nysius wekiuicola*) is a unique bug found only on Mauna Kea above the 11,715-foot (3,571-meter) level. "*Wēkiu,*" a Hawaiian word meaning "top" or "summit," refers to the bug's habitat. The tiny, quarter-inch-long *wēkiu* lives under large boulders and in loose cinder, which protects it from cold, allows it to get moisture, and otherwise helps protect it from harsh conditions at the summit.

The *wēkiu* is interesting because of an adaptation it has made to its living conditions. While most bugs are herbivores, eating herbs and plant juices, the *wēkiu* has adapted to its harsh environment by becoming a predator. Using its straw-like mouth structures, it eats wind-blown insects swept up to the summit of Mauna Kea from surrounding lowlands. These insects, not adapted to the summit's extreme conditions, hide in protected areas such as beneath cinder and rock, where they are easy targets for the *wēkiu. Wēkiu* are thought to eat ladybugs, flies, moths and butterflies. It's also thought that *wēkiu* bugs might have glycerol-like compounds in their blood that keep their bodies from forming ice crystals in subfreezing temperature, giving them an advantage over the insects they eat.

Most *wēkiu* live on the inner slopes of *pu'u* craters, which are made of loose volcanic cinder and provide protective cover as well as protection from the sun. The *wēkiu*'s habitat is near areas of snow, suggesting that the bug might also, at times, depend upon slowly melting snow for moisture.

Wēkiu numbers have been declining—the population is thought to have shrunk significantly between 1982 and 1999—causing the bug to be designated as a candidate for listing as an endangered species. It's not certain why the numbers are declining, though possible reasons include human impact and climate change, among others. One *wēkiu* habitat is Pu'uhau'oki, near the Keck Observatory. Concerns about the well-being of the *wēkiu* during construction of the

The nēnē, *Hawai'i's state bird, lives in Mauna Kea's* māmane *forest.*

Keck Observatory's new "outrigger" telescopes have led to the establishment of a *Wēkiu* Bug Mitigation Plan. The plan calls for the bugs' numbers to be monitored and for construction to be carefully conducted so as not to impact the *wēkiu* habitat or inadvertently introduce other creatures that might interfere with the survival of the *wēkiu*. Note that all cinder slopes on the mountain are *wēkiu* habitats, which is another reason for visitors to stay on established trails.

PALILA

In the open woodland areas of Mauna Kea between 6,500 and 10,000 feet (1,981 and 3,048 meters) live numerous types of birds, including the *palila* (*Loxioides bailleui*) and the seabird *'ua'u* (*Pterodroma phaeopygia sandwichensis;* Hawaiian petrel), both endangered. The *nēnē*, a Hawaiian goose and the state bird, also lives in the *māmane* forest.

The *palila* is a small, finch-billed Hawaiian honeycreeper with a big bill, a large yellow head and throat, and a grayish back and lighter underside. *Palila* were once widespread, living on the Big Island from Mauna Kea to Kona as well as on O'ahu and Kaua'i. Now they are found only on Mauna Kea, and they are the last remaining Hawaiian bird to live only in dry forests. They have been listed as endangered since 1966.

The birds have a unique relationship with the mountain's *māmane* trees. They eat *māmane* leaves and flowers, and use their bills to open *māmane* seedpods and eat the tender green seeds. Along with moth caterpillars, which live inside the seedpods, the *māmane* makes up most of the *palila*'s diet. Over time, the *māmane* developed a toxin to protect their seeds, but the *palila* adapted, too, as did caterpillars in that habitat, developing a resistance to the toxin. The *palila* feed protein-rich moth caterpillars to their young rather than *māmane* seeds, probably because the caterpil-

The māmane *forest, apart from its beauty, is a critical habitat of the* palila.

lars don't have the toxin that the seeds do.

Māmane trees are also important to the *palila* for shelter and cover; the birds build their nests in the trees and hide their young from predators and the sun. More than 30,000 acres of *māmane* and *naio* forest on Mauna Kea have been designated a critical habitat of the *palila*.

Though the *palila* are endangered, scientists say there is an excellent chance they will survive and perhaps even return to some of their previous nesting grounds. Because they live high up Mauna Kea, they are not subject to some of

This painting of a palila *shows it feasting on* māmane *seedpods.*

the avian diseases that strike other bird populations. Mosquitoes are not a threat, because standing water does not accumulate in the sub-alpine forests where the birds live. Also, *palila* numbers are not so low that they are creating the genetic inbreeding problems that can occur when the breeding pool is exceptionally small.

Another endangered species known to be living on Mauna Kea is the *'io* (*Buteo solitarius*; Hawaiian hawk). Other bird species on the mountain include *'i'iwi*, *'amakihi*, *'apapane*, *'elepaio* and *pueo*, as well as game birds such as turkey, California quail, Chukar partridge, and Erckel's Francolin.

MAMMALS

Hawai'i's only surviving native land mammal is the endangered Hawaiian bat, *'ōpe'ape'a* (*Lasiurus cinereus*). The *'ōpe'ape'a* has been seen in the *māmane* forests of Mauna Kea.

Cattle were introduced to the saddle area between Mauna Kea and Mauna Loa in the late 1700s and early 1800s where they soon ran wild, as did feral sheep and goats. In 1873, Isabella Bird wrote about visiting a sheep station at Kalaieha on the southern slope of Mauna Kea, where there were 9,000 sheep.

Mauna Kea's feral animal population is made up of animals that were let loose to multiply on the mountain or that escaped from domestic herds, which in turn descended from those that arrived with the earliest European explorers and merchants, most before 1800 (though pigs arrived in Hawai'i much earlier, with the first Polynesians).

As the animals' numbers increased, Mauna Kea's vegetation underwent a serious decline. Feral sheep and goats ate *māmane* leaves, stems, seedlings and sprouts, destroying a great deal of that ecosystem. By 1850, cattle had eaten a significant portion of the upper *'ōhi'a-koa* forest. Much of the mountain's vegetation was gone by the late 1920s, when the government instituted an eradication program.

Cattle were eliminated from the mountain, and feral sheep numbers declined from about 40,000 in the mid-1930s to only about 200 two decades later. In areas where hunters decreased the game populations, forests began to regenerate. In less accessible areas, animals continued to damage the fragile forest, including the *palila* habitat.

In 1979, a federal court order mandated the removal of feral sheep and goats from more than 30,000 acres of *māmane* and *naio* forest. In 1987, another order called for the removal of mouflon and hybrid sheep. The resulting eradication programs have made a difference—some areas have experienced a significant regeneration of their vegetation. ▲

RECREATION 4

4 Recreation

Visitors often ascend Mauna Kea just to play in the snow.

If you plan to ski, snowboard, hike or hunt on the slopes of Mauna Kea, always remember where you are. Mauna Kea is a place with significant spiritual and cultural importance to a great number of people. Treat the land, as well as the rocks and plants and everything else you come upon, with respect. Never drive "off-road"—many cultural and archaeological sites on the mountain are not obvious to the untrained eye, and you may unintentionally cause irreparable damage to those or other natural resources.

Safety is also a concern when exerting yourself at Mauna Kea's high altitude. Take responsibility for your own well-being and be responsible about going up in the thinner oxygen (see tips on preparing in the section "Altitude and Safety," p. 13), and about what you do on the mountain. Keep in mind, too, that the only facilities available above the Visitor Information Station are portable toilets at Parking Area #3 and near the UH 0.6-meter Telescope, and the Keck Observatories' rest-rooms, which are open on a limited schedule (weekdays from 10 a.m. to 4 p.m.).

PLAYING IN THE SNOW

You know it's winter in Hawai'i when a pickup truck, its bed full of snow, rushes down Waiānuenue Avenue, and warm Hilo weather causes icy water to stream from its tailgate. Some of that snow, hauled from the heights of Mauna Kea, is dumped in green yards where tan, barefoot people wearing shorts and

Snow is packed in a pickup truck to astonish people below.

tank tops pack it into unlikely tropical snowmen next to flowering plumeria trees and *hāpu'u* (tree ferns) full of brilliant blooming orchids.

Some people bring snow down the mountain in coolers more commonly used for beach trips, and then dump the packed white rectangles on lush green grass for babies to touch and dogs to taste.

Typically, snow lasts longest on the northern slope of Pu'uhaukea (Hill of White Snow); sometimes that's the only place on Mauna Kea with snow.

February and March are the months you're most likely to find snow on the mountain, though once in a while it stays all through the summer. There is more precipitation, and therefore more snow, in La Niña years; El Niño years are drier.

People heading up the mountain to play in the snow generally park along the roadways and play nearby. The Mauna Kea Observatory Support Services keeps the summit road cleared of snow.

Hikers test the chilly waters of Lake Waiau.

Hiking is more strenuous on Mauna Kea due to the high altitude.

Remember that there is no food or water (except when the Visitor Information Station is open), and gas and towing services are not available to visitors on the mountain. Come prepared. And dress for the cold.

HIKING

In pre-Western-contact times, people probably climbed Mauna Kea by following ridgelines or water sources. We don't know what trails they used in those days, though we know from oral tradition and other family stories that Hawaiians have long journeyed up the mountain for cultural, spiritual and other reasons. As they hiked, early Hawaiians probably visited shrines their families had erected to their gods. Some still do.

In the late 19th and early 20th centuries, people often ascended the mountain on horseback. Trails heading *mauka* (upland) led up from most of the different *ahupua'a* (pie-shaped land divisions stretching from the sea to the mountain-top) along Mauna Kea's flanks. The trails generally rose from ranch stations along the mountain slopes where there were accommodations and provisions for horses.

Today there are a few trails on Mauna Kea, but they are unmarked, unmaintained, and without facilities. Stay on the trails! Check in at the Visitor Information Station or with a Ranger before and after you hike. Hikers should keep in mind that hiking is much more strenuous on the mountain, especially near the summit, because of the high altitude. At this altitude, the air is cold and the sun's light is especially strong: bring sunglasses, sunblock, warm clothing, and your own food and water. Always pack out your trash. Remember, too, that the weather can change suddenly and drastically.

That said, it's a beautiful place to hike. The skies are clear, the air is cool and the landscape can be stunning. Most people drive up the mountain and park, then hike along existing roads in the Ice Age Natural Area Reserve. Day hikes are allowed in the Natural Area Reserve, though groups of 12 or more need a permit from the Department of Land and Natural Resources. Commercial groups are not permitted.

One trail, the Humu'ula-Mauna Kea trail, runs from an old sheep station at Humu'ula to Lake Waiau in the Natural Area Reserve. Most people who went up Mauna Kea before 1960, when the summit alignment trail was opened, used this trail.

HUNTING ON MAUNA KEA

Hunting, an old tradition in Hawai'i, has long occurred on Mauna Kea, and it continues on the mountain's lower slopes. A sport for some, it's an important part of the culture and lifestyle and a means of subsistence for others—and it can help to control encroaching feral animal populations.

If you haven't hunted in Hawai'i before, and/or are not familiar with current hunting regulations, call the Waimea Fish and Wildlife office at 887-6063, or the Hilo office at 974-4221, before heading out.

Here hikers are alone with the mountain and the sky.

There are three hunting units on Mauna Kea. Unit K is in the Ice Age Area, and along with Unit A it is known as the Mauna Kea Game Management Area. These two are operated as one area with the same rules. Both units are open for sheep and pigs every day of the year (except for bird-hunting days; see below). There is no "bag limit" on sheep, and hunters may take one pig per person per day. These units are rifle-hunting areas; dogs are not allowed for mammal hunting.

Unit G, the Ka'ohe Game Management Area, is an archery-hunting area. It, too, is open every day of the year, with no bag limit on sheep and a limit of one pig per hunter per day. Muzzle-loading rifles are allowed in Unit G for sheep hunting, and therefore blaze-orange vests are required in this area (they are not required in other archery areas). No dogs are allowed.

From March through mid-April it's spring turkey season in Units A and G. Shotguns and archery are allowed, though no dogs, and tags are required. The bag limit is usually two male turkeys per day and no more than three per season.

Game bird season runs from the first Saturday in November through the third weekend in January (including Martin Luther King, Jr. Day). Bird hunting is on Saturdays, Sundays, Wednesdays, and state holidays in both areas. This may vary due to the number of birds available. Check with the Fish and Wildlife Office for current information.

See *http://www.state.hi.us/dlnr/dofaw/rulesindex.html* for hunting rules (download chapter 122 for regulations regarding bird hunting and chapter 123 for those about game mammals).

Nowadays there are limits for most hunting on Mauna Kea.

HUNTING LICENSE INFORMATION

If you're planning to hunt while visiting Hawai'i, it's crucial that you plan ahead. Many hunters from out-of-state end up having to cancel their plans because they arrive without the proper documents.

A hunting license is required to hunt in Hawai'i, whether on public or private land. To get a license, you are required to have either a Hawai'i Hunter Education Wallet Card or a Letter of Exemption.

The Hunter Education Wallet Card is obtained upon graduating from the basic Department of Land and Natural Resources hunting course.

Letters of Exemption are issued free to those who have an out-of-state hunter education card or a Hawai'i Hunting License issued prior to July 1, 1990. Apply for a Letter of Exemption online at *http://www.ehawaiigov.org/DLNR/hunting*

If you're not a Hawai'i resident, and don't have a hunter education card from your home state, contact your local hunter education office to find out how to enroll in a hunter education course.

If you have either a Hawai'i Hunter Education Wallet Card or Letter of Exemption, you're set to get your hunting license, which you can buy online at *http://www.ehawaiigov.org/DLNR/hunting*

For more information, check out the DLNR's website at *http://www.state.hi.us/dlnr* or contact them at Department of Land and Natural Resources, Division of Forestry and Wildlife, 1151 Punchbowl Street, Room 325, Honolulu, HI 96813. Phone 587-0166; fax 587-0160.

The Visitor Information Station

5

5 The Visitor Information Station

THE GATEWAY TO MAUNA KEA

The Mauna Kea Visitor Information Station (VIS) is located at the 9,200-foot (2,804-meter) level of Mauna Kea. It is strongly recommended that you stop at the VIS to acclimate to the high altitude before continuing to the summit. Here, you can also read about health and safety issues and learn about the cultural, natural, historical and astronomical treasures of Mauna Kea. Spending one hour at the VIS before continuing to the summit helps to reduce the chances of altitude sickness.

Here is the Visitor Information Station logo.

The VIS is located within the Onizuka Center for International Astronomy, which includes the entire astronomical support facility at the 9,200-foot (2,804-meter) level. It is often referred to as Hale Pōhaku ("House of Stone"). The name refers to the original stone edifices built at this site in the 1930s and still standing today.

Don't confuse the VIS with the Astronaut Ellison S. Onizuka Space Center at Keāhole Airport in Kailua-Kona or the Kīlauea Visitor Center at Hawai'i Volcanoes National Park—those are different facilities, and both are long drives

Did You Know?

Mauna Kea's Visitor Information Station is widely considered the best place on earth to view the night sky with the unaided eye or amateur telescopes.

Did You Know?

- Open every day of the year: 9 a.m. to 10 p.m.
- Free stargazing program every night of the year: 6 p.m. to 10 p.m.
- One-hour drive from Hilo and Waimea; two-hour drive from Kailua-Kona and Hawai'i Volcanoes National Park
- Website: http://www.ifa.hawaii.edu/info/vis

from Mauna Kea if you end up at the wrong place!

The station is open every day of the year from 9 a.m. to 10 p.m., and the public restrooms are open 24 hours a day. There are friendly and helpful interpretive guides, student staff, volunteers and Rangers available to answer questions and inform you about Mauna Kea. The VIS also has an excellent selection of interpretive panels and interactive computers and a wide range of educational videos that you can request.

As you enter, notice the flags near the entrance. These flags represent the nations and states with interests in the various telescopes on Mauna Kea: the United States, Hawai'i, the United Kingdom, Japan, Canada, France, the Netherlands, Taiwan, Chile, Brazil, Australia and Argentina.

Educational products, convenience-style food, hot cocoa and warm clothing can be purchased at the VIS's First Light Bookstore. The station also has a variety of programs, all free to the general public. A donation box is located on the handout table; all donations are tax-deductible (University of Hawai'i Foundation, Account No. 12045164). Be sure to dress warmly, as temperatures can be near freezing.

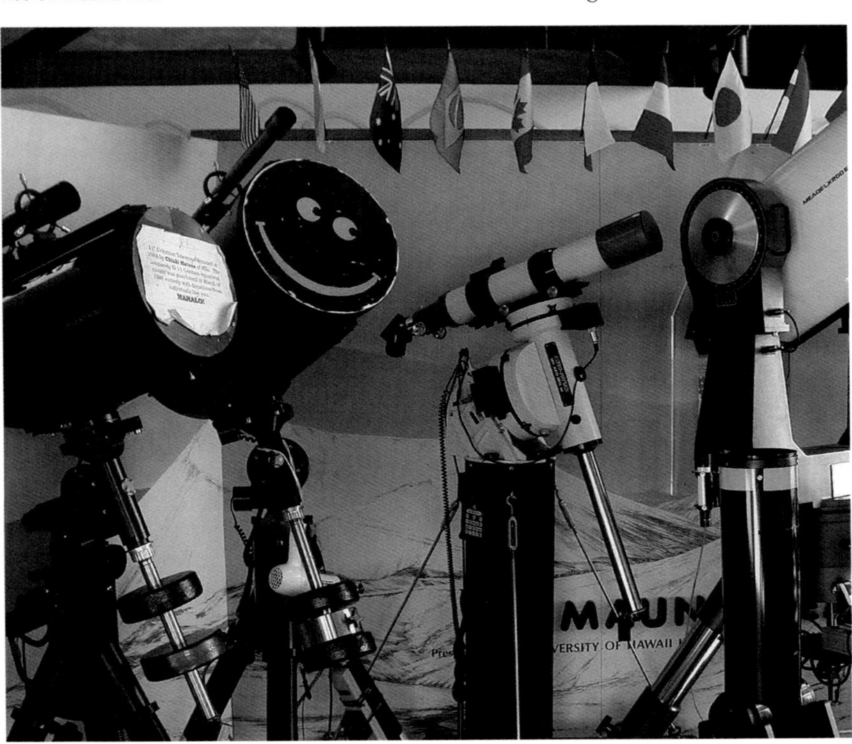

Eleven nations have interests in the observatories on Mauna Kea.

ELLISON SHOJI ONIZUKA MEMORIAL

A memorial to astronaut Ellison Onizuka, who was born and attended school on the island of Hawai'i, is located on the west side of the Observation Patio. Onizuka was serving as a Mission Specialist aboard the space shuttle *Challenger* when it exploded on January 28, 1986. He held the rank of lieutenant colonel in the United States Air Force and earned bachelor's and master's degrees in aerospace engineering from the University of Colorado. He served as a test pilot and as a flight test engineer when he entered active duty. Onizuka was accepted for the astronaut program in January of 1978 and also flew on board the space shuttle *Discovery* in 1985.

A memorial honors Ellison Shoji Onizuka, who died in the space shuttle Challenger *accident.*

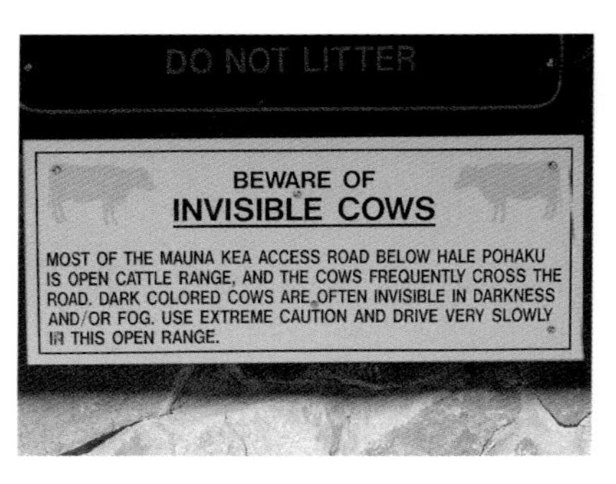

Watch out for "invisible cows" in the fog!

OUTSIDE INFORMATION DISPLAYS

Before entering the front doors of the station, note the written panels that provide basic health and safety information along with advice on proper conduct while on the mountain. Be sure to read about "invisible cows"! From time to time, cows escape from the fenced pastureland downhill from the VIS. Many of them blend into the fog and seem to become invisible. Be careful driving in this area—some of these cows, while "invisible," weigh more than 2,000 pounds (907 kilograms)!

BROCHURES

There are many free brochures available at the station. Be sure to pick up "Mauna Kea Hazards," "Visiting Mauna Kea, Safety and Responsibility," and the pamphlet on culture and Mauna Kea. Taking a minute to read through these could prevent a serious problem during your visit. There are also a variety of brochures, handouts and fact sheets provided by the observatories. Each contains more than enough information for a typical high school report or project, and they are a great source for "What I Did during My Summer Vacation" reports at any grade level. Don't forget to fill out the program evaluation form, which helps the VIS improve programs and displays.

A number of free, public stargazing programs are offered at the VIS.

STARGAZING PROGRAM

This program is offered every night of the year from 6 p.m. to 10 p.m. and begins with a short introduction and a one-hour video about Mauna Kea. View the most beautiful objects in the night sky from the best stargazing site on earth! Typically, six telescopes, ranging from a 16" Schmidt-Cassegrain to 4" refractors, are placed on the Observation Patio. Interpretive guides, astronomy students from the University of Hawai'i at Hilo, and volunteers

Visitors at the VIS can study both daytime and nighttime stars.

operate the station's telescopes and explain the various celestial bodies. You can see many objects in the southern sky not typically visible from the U.S. mainland. The station also has several telescopes available for visitors to operate. Take some time to use one of these and you can truthfully say that you operated a Mauna Kea telescope.

Solar telescopes reveal sunspots, solar flares and prominences.

SOLARGAZING PROGRAM

This program offers a variety of solar telescopes and instruments, which are available every day from 10 a.m. to 6 p.m. Using the instruments at the station one can see sunspots, solar flares, prominences and the beautiful solar spectra with absorption lines. You can even make a diagram of the sun with sunspots and stamp it with an "Official Mauna Kea Data" stamp.

ESCORTED SUMMIT TOURS

Every Saturday and Sunday, a free escorted summit tour meets at the station at 1 p.m. This is the best way for the general public to see inside the observatories. The tour convoys to the summit, where it typically visits the Keck Observatory Visitor's Gallery and University of Hawai'i's 2.2-meter Telescope, then returns to the station by 5 p.m. Alternately, you can opt to stay on the summit and watch the sunset. Please leave the summit area within a half-hour after sunset, so your headlights and engine heat will not interfere with the observatories. Requirements for participation in the escorted summit tour are:

- You must have a four-wheel-drive or all-wheel-drive vehicle.
- Everyone in the vehicle must be at least 16 years old.
- Everyone in the vehicle must be in good health and without any cardiovascular problems.
- There must be no one in the vehicle who is pregnant.
- There must be no one in the vehicle who has been scuba diving in the past 24 hours.

THE UNIVERSE TONIGHT

The Universe Tonight is presented on the first Saturday of every month at 6 p.m. Each month a different Mauna Kea astronomer presents new discoveries and current research at his or her observatory. After the hour-long presentation, the astronomer conducts a question-and-answer period, and then joins the stargazing program at the VIS, when you'll have an opportunity for one-on-one discussion. This is a great way to talk to a research astronomer and hear about current scientific discoveries and maybe even learn about an unannounced finding.

MA LALO O KA PŌ LANI

Ma Lalo O Ka Pō Lani ("Under the Night Sky") is a program on Hawaiian culture, presented the third Saturday of every month at 6 p.m. for about an hour preceding the stargazing program. Each month a native Hawaiian cultural practitioner speaks on a cultural issue related to Mauna Kea.

Sunset colors the sky at the VIS.

METEOR SHOWERS AND SPECIAL ASTRONOMICAL EVENTS

The station remains open all night for the Perseid meteor shower that occurs every year between August 12 and 15 and for the Leonids that occur between November 13 and 19. There are often a few thousand visitors at these programs. The Hilo Astronomy Club, the Astronomy Club at the University of Hawai'i at Hilo, and the VIS volunteer corps organize these all-night programs. The station also schedules programs for special astronomical events such as eclipses, transits and occultations. Call the station

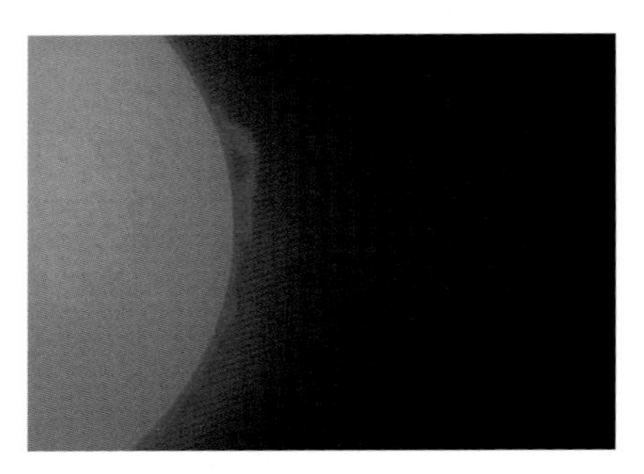

Solar flares are shown here erupting off the sun's surface.

at 961-2180 or visit its website (*http://www.ifa.hawaii.edu/info/vis*) for specific events, dates and times, as well as links to Mauna Kea observatories and other astronomical sites.

ASTRONOMY CLUB AT THE UNIVERSITY OF HAWAI'I AT HILO NIGHT

Members of the Astronomy Club at the University of Hawai'i at Hilo conduct the stargazing program and operate the station on the second Saturday of the month during the fall and spring semesters. This is a popular program during which the general public, especially children and teenagers, can interact directly with astronomy students.

HILO ASTRONOMY CLUB

On the Saturday closest to the new moon, the Hilo Astronomy Club participates in the station's stargazing program. These are the best nights to view deep-sky objects such as galaxies, nebulae and star clusters. Up to 15 telescopes are set up on the station's Observation Patio on these nights. Club members share their telescopes and expertise with all who show up—amateur astronomers are welcome on these nights, too. So whether you have a telescope, a pair of binoculars or just your eyes, come and join the club!

OBSERVATION PATIO

The Observation Patio is open to the general public 24 hours a day. Amateur astronomers are encouraged to bring their telescopes and observe all night long. Bathrooms, electricity, drinking water and a public telephone are available all night. This is a great way to spend a night in Hawai'i, under the best night sky in the world. Note that there's a switch on the streetlight pole that turns the parking lot lights on and off.

SPECIAL REQUESTS

Groups may submit a request

A visitor poses at the Ellison Onizuka Memorial outside the VIS.

Interpretive displays help make a visit to the VIS educational as well as fun.

for specific programs, including stargazing, escorted summit tours, cultural programs, natural history walks and other special programs. Qualified groups include the following:

- Bonafide educational institutions
- Community service groups
- Groups associated with culture, natural history, or astronomy

Download a special request form from the station's website and fax it back at least a month before the proposed date. Reply will be via fax, and, if the request is approved, there may be a fee. Groups with an approved special request can either drive themselves or be transported by one of the commercial tour companies permitted to take groups up the mountain. Groups must make the special request directly to the VIS, though. Commercial vendors cannot do so, as they are not eligible for special requests.

Presentation and activity rooms seating up to 75 people are also available to qualified groups for lectures, training, demonstrations and other activities.

FILM PERMITS

Commercial film groups and professional photographers wishing to take photographs on the mountain must acquire a permit from the Office of Mauna Kea Management (OMKM). There may be a fee,

The First Light Bookstore displays space-themed food and other products.

including the cost of a Mauna Kea Ranger to accompany the group. Contact the OMKM at 933-0734.

FIRST LIGHT BOOKSTORE

The First Light Bookstore is open at the VIS every day from 9 a.m. to 9:30 p.m. It offers a variety of educational products, convenience-style food and hot cocoa. Be sure to check out the candy bars and cookies—there's a wide selection with a decidedly astronomical theme. Popular gifts include T-shirts with Hawaiian star names and replicas of the benchmark, a surveying marker on the summit. Warm clothing can also be purchased. The bookstore carries a wide selection of books related to astronomy and natural history, as well as several types of star charts. In fact, the First Light Bookstore is the only place in the Milky Way Galaxy where many of its unique products are offered. All proceeds support the VIS's educational programs.

VOLUNTEER PROGRAM

There are over 100 volunteers at the VIS, many of them astronomy students from the University of Hawai'i at Hilo. They contribute more than 9,000 hours per year and lead programs such as stargazing and escorted summit tours.

Some of the student volunteers have completed projects using the large telescopes on the summit and are knowledgeable about current astronomical discoveries. Young adults, teenagers and children readily identify with the student volunteers—point your children toward the volunteers to hear about their experiences at the university and on Mauna Kea, and your kids may become inspired to study astronomy, perhaps even to become astronomers or astronauts.

MAUNA KEA RANGERS

There is always at least one Mauna Kea Ranger on duty on the mountain. They are well trained and available to render aid, answer questions and protect the cultural, natural and astronomical resources of Mauna Kea. Feel free to ask for assistance in case of accident or illness, especially altitude sickness. Rangers can usually be found at the VIS or on summit patrol—look for vehicles clearly marked "Ranger." Remember that Mauna Kea is a remote location where extreme weather can occur at any time and without notice, so it's important to comply with all instructions from the Rangers.

Sunrise and sunset are equally beautiful at the mountain's summit.

HIKES AT THE STATION

There are several short hikes in the vicinity of the station that take from 15 minutes to two hours. The four principal hikes are:

Silversword Enclosure—The wooden gate at the back of the Station's parking lot is the entrance to a silversword enclosure officially known as the Hale Pōhaku enclosure. It takes about 20 minutes to walk though this area. There are between 25 and 50 Mauna Kea silverswords here. As you walk through, be sure to observe the *kuahu lele*, a modern, traditional native Hawaiian altar. Ceremonies are conducted here from time to time; please look, but do not touch.

Kilohana—The walk to the Kilohana picnic tables is actually the beginning of the Mauna Kea Humuʻula Trail, which goes all the way to the summit, and it gives you a taste of the long and challenging summit hike.

West Ridge—This is a short walk to the ridge directly across the road from the station. There is a good view of Puʻu Haiwahine from this vantage point.

Puʻukalepeamoa—The best place to view the sunset within walking distance from the station

The gorgeous colors of a Big Island sunset reflect off the snow.

is from this high peak. This cinder cone across the road from the VIS provides a great vantage point from which to see Mauna Loa and the west side of the Saddle. You'll need to dress warmly and carry a flashlight for the return trip if you hike up for the sunset.

SUNSETS

Besides the Pu'ukalepeamoa hike, other good places to see great sunsets are the sharp turn about a third of a mile below the VIS and numerous vantage points along the road to the summit. There are other vehicles traveling on these roads, so be sure to choose a location where they can pass safely.

A TYPICAL FAMILY DAY ON MAUNA KEA

Saturday is the best day to bring the family to Mauna Kea, as there are many activities to keep everyone interested. A typical family trip to Mauna Kea might look like this:

Departure—11 a.m. from downtown Hilo, or 10 a.m. from

downtown Kailua-Kona.

12 noon—Arrive at the VIS, visit the silversword enclosure and have lunch.

1 p.m.—Escorted summit tour begins. If you are younger than 16 years or for other reasons don't plan to go on the escorted summit tour, you can take a hike near the VIS, use one of the interactive computer terminals, watch a video, make diagrams of sunspots, view the sun through the station's three solar instruments or shop at the First Light Bookstore.

5 p.m.—The escorted summit tour returns to the VIS. However, very few visitors return with the escorted summit tour—most stay on the summit to watch the sunset. The sun sets at different times depending on the time of year; your guide will give you the current time. If you choose to stay, let the interpretive guide know. After sunset, join the stargazing program down at the VIS.

6 p.m.—Stargazing program begins.

10 p.m.—Stargazing program ends and the station closes.

CELL PHONES

Cell phone usage on the summit area is discouraged except for emergencies. The radio transmissions from the cell phone may interfere with some of the telescopes. ▲

A visitor views a partial solar eclipse through protective goggles.

6

Astronomy on Mauna Kea

6 Astronomy on Mauna Kea

MAUNA KEA—A UNIQUE MOUNTAIN FOR ASTRONOMY

Mauna Kea is widely considered the best site on earth to study the universe with research telescopes. This is due to many factors, both environmental and civil:

Height of Mauna Kea—At 13,796 feet (4,205 meters) elevation, the summit region is above close to 40 percent of the earth's atmosphere. This means the degradation of starlight caused by the atmosphere is significantly reduced.

Dry Air—The air above the summit is very dry, which is important for infrared and submillimeter observations. Water vapor absorbs these wavelengths of radiation. There are only a few sites on earth dry enough to make these types of observations.

Cloud-Free—The number of cloud-free nights per year above Mauna Kea is among the highest in the world.

Stability of the Atmosphere—The extraordinarily stable atmosphere above Mauna Kea produces less distortion; therefore, more detailed studies are possible here than elsewhere.

Tropical Inversion— A tropical inversion layer typically exists at about 8,000 to 10,000 feet (2,438 to 3,048 meters) elevation. It's about 2,000 feet (610 meters) thick, and it isolates the summit region from the moist maritime air below. The inversion also keeps particulates such as dust and volcanic gases from rising to the observatories.

Dark Sky—Mauna Kea's summit is located far from any city lights, and the dark sky and a strong island-wide lighting ordinance ensure that the site will retain its integrity for the foreseeable future. (Lights from nearby population centers interfere with major observatories on the mainland U.S.)

Latitude—The island of Hawai'i is located at about 20 degrees north latitude. Because of

Mauna Kea is widely considered the best stargazing spot on earth.

this location, all of the northern sky and much of the southern sky is visible.

Geographical Location— The location of Mauna Kea in the central Pacific Ocean results in a clear, stable atmosphere. The constant, regular tradewinds minimize turbulence in the lower atmosphere.

Shape of Mauna Kea—The symmetrical shape of the mountain results in laminar (smooth) airflow around and over the mountain, which decreases the amount of mixing and therefore turbulence in the air and results in less distortion.

Modern Infrastructure— A modern and technologically advanced infrastructure is essential to support a world-class observatory complex. This includes ready access to deep-draft harbors and wide-body airports, a modern communications system and a stable government. Community-based facilities are also necessary to retain a highly educated staff, including diverse educational opportunities, modern medical facilities, good housing and an active social life. All these are available on the island of Hawai'i.

Base Support Facilities— Each observatory has a base support facility, located either in University Park at the University of Hawai'i at Hilo or in the ranching town of Waimea. Base support facilities house the administrative staff, transportation facilities, computer networks, and development groups for computer programming and instrumentation development. Many of the telescopes can be controlled from and receive data at the base support facilities, allowing visiting astronomers to work at sea level and reducing the number of staff needed on the summit during observations.

A SELF-GUIDED DRIVE TO THE SUMMIT

A self-guided tour to the summit begins at the Visitor Information Station. Be sure to heed health and safety recommendations and rules:

- Spend a minimum of a half-hour—preferably an hour—at the VIS.
- A four-wheel- or all-wheel-drive vehicle is strongly recommended.
- Visitors going beyond the VIS should be 16 years old or older.
- Visitors to the summit should be healthy, with no cardio-pulmonary issues.
- Pregnant women must not travel to the summit.
- If you have been scuba diving in the past 24 hours, you should not go to the summit.

Before you begin, be sure to do the following:

- Use the bathrooms at the VIS. The portable facilities further on are very basic.
- Drink water, as it can be very dry at the summit.
- Carry plenty of water. The

dryness can cause rapid dehydration.

- Dress warmly: it is almost always near-freezing on the summit, and the two visitor's galleries are refrigerated.
- Wear sunscreen—anyone, regardless of their complexion, can get severe sunburn, not to mention skin cancer, due to the high levels of ultraviolet radiation.
- Wear sunglasses, as the ultraviolet radiation can also cause snow blindness and premature cataracts.

Prepare your vehicle:

- Open and close your gas tank cap; this equalizes the air pressure inside the tank.
- Make sure that your gas tank is nearly full before leaving town (at least half-full before leaving the VIS), as gasoline is not available on Mauna Kea.
- Put your vehicle in four-wheel-drive, low-range.
- Wear your seat belt—it can be a bumpy ride.
- Reset your trip counter—this will help you locate sites en route to the summit.

A good time to begin a tour from the VIS is about two hours before sunset. This gives you plenty of time to drive up to the summit with stops along the way, see the observatories and find a good spot to watch the sunset. The time the sun sets varies throughout the year, so check before you leave. If you wish to go inside the three observatories with public galleries/tours, Subaru, the Keck I and/or the University of Hawai'i 2.2-meter Telescope, check their opening and closing times and plan accordingly. You'll need reservations for the Subaru tour—see the Subaru Telescope section (pp. 104-105) for details.

The road from the VIS to the summit is about eight miles long, with paving that stops just above the station and starts again below the 5.0-mile marker. There are many switchbacks and turnarounds along the road, and the steepest slope has a grade in excess of 15 percent. Twelve of the 13 telescopes are located near the summit and lie in a roughly circular loop.

As you leave the station, notice the chalet-type buildings above the VIS and to your right. This is Hale Pōhaku, where astronomers and telescope operators sleep during the daytime after observing the previous night. If you think of a typical college dorm room and imagine it cut in half, that's about how much space the astronomers have. There is also a cafeteria, observatory offices and living areas. These facilities are available only to active observatory staff and observers.

Also notice the snowplows, snow blowers and other pieces of heavy equipment. You probably never thought you would see snowplows in Hawai'i, but winter storms often bring several feet of snow.

Mile marker 0 ("0.0 MM" on the map on page 10) is where the paving stops and the gravel road begins. Drive especially carefully for the next half-mile. The first switchback is dangerous due to its steep sides, lack of road shoulders and narrow width.

Astronomers and telescope operators sleep in these quaint dorms—during the day.

Just before you reach mile marker 1.0, look back downhill and you can see the entire Onizuka Center for International Astronomy complex (the VIS, Hale Pōhaku, the Utilities Department and the Construction Camp). On most days you can see all the way to the Saddle Road. Continue to the second switchback, at about 1.5 miles, where there is a lookout. You can see most of the western saddle area of the Big Island from here. As you pass mile marker 2.0, take another look at the Onizuka Center and the Saddle Road. The view gets better the higher you go. Stop at the turnaround at the third switchback, at about 2.7 miles. In addition to the great view, there is also an area where silverswords grow (please don't touch them).

As you pass MM 4.0, notice Pu'u Keonehehe'e, the large black cinder cone on your left. At about 4.4 miles you are actually driving on a glacial terminal moraine; notice the glacial till on your right. This area, sometimes called Moon Valley, is where the Apollo astronauts trained during the 1960s with their "Moon Buggy."

Paving starts again at about the 4.7-mile point. Keep your vehicle in four-wheel drive, low range, and resist the temptation to drive faster than the posted speed limit. Remember that the low oxygen at higher altitudes impairs your judgment and slows your reflexes. You'll pass Parking Area #1, a small parking lot sometimes used for emergency medical evacuations. The entrance to the Very Long Baseline Array is at about the 5.3-mile point.

Pu'u Ko'oko'olau and Keanakāko'i, the ancient adze quarry, are to your left at about 5.5 miles. You can tell where Hawaiians quarried for basalt by noting the different coloration in the cliff faces. Please don't walk out to the quarry; any traffic in this very fragile area can damage it.

Pass MM 6.0 and continue to Parking Area #2 at almost 13,000 feet (3,962 meters). The trailhead to Lake Waiau is across the road, though the lake is not visible from there. To access the trail to the lake from Parking Area #2, walk down the road until you see a signpost marking the beginning of the trail before beginning to traverse the terrain. (Footprints can last a long time

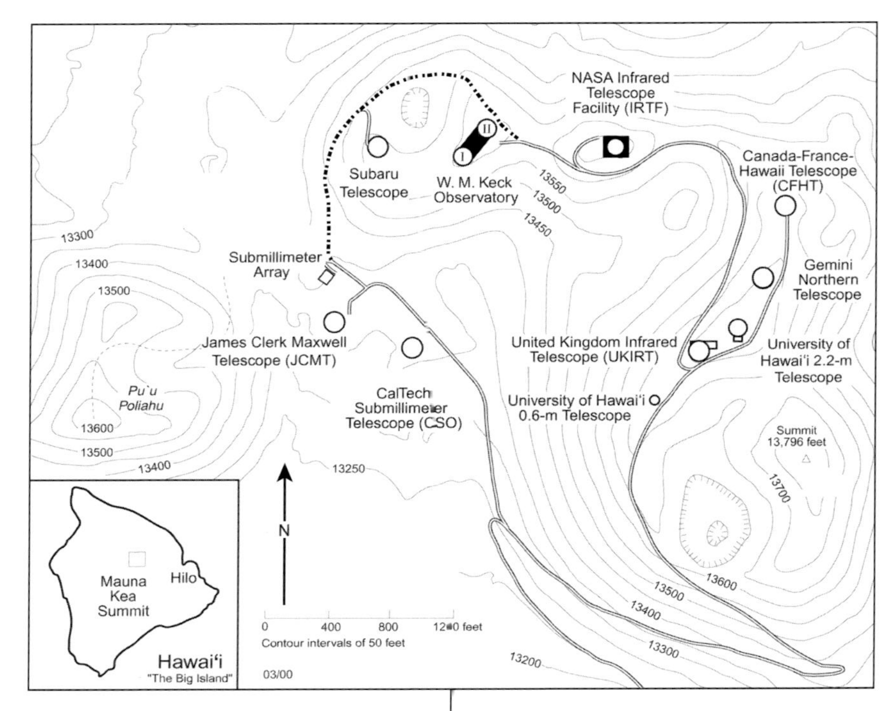

A topographical map gives the locations of the Mauna Kea observatories.

and encourage others to walk off-trail.) You may also access another trail to Lake Waiau from Parking Area #3.

Continuing up the road, you enter what is commonly known as Goodrich Pass. Be very careful, especially in the early morning, as black ice forms very quickly on this steep, shaded section of the road, and it's easy to lose control of your vehicle. Beautiful Pu'uhaukea is on your left and Pu'u Wēkiu on your right.

Just above MM 7.0 is the first summit switchback, as well as Parking Area #3 and three portable toilets. You'll see the first summit observatories to your left. All the submillimeter telescopes are located in this area, which is commonly called "Submillimeter Valley." At this point, turn left to follow this guide.

First you'll pass the silver, spherical Caltech Submillimeter Observatory on your left. Farther up the road is the James Clerk Maxwell Telescope; you'll recognize it by its large white carousel and flat roof.

Farther along is the Submillimeter Array, again on your left. Notice the futuristic antennas that make up the elements of this array. It is common to see Haleakalā on Maui off to your left as you travel along this gravel road. There is a 5 mph speed limit on the cinder road above SMA because the dust created damages the antennas and receivers.

The entrance to the Subaru Observatory is located approximately 300 yards (274 meters) beyond the end of the pavement. Drive slowly, 5 mph, on this unpaved section of the road. Notice the large, silver, rectangular dome and the large support building next to it.

Continuing, you'll come to a "T" in the road. Make a right-hand turn and you will be in front of the W. M. Keck Observatory with its twin telescopes. Keck I is on your left and Keck II on your right as you face the observatory from the parking area. Keck I has a visitor's area that is open Monday through Friday from 10 a.m. to 4 p.m. There are clean bathrooms in the visitor's area, as well as a drinking fountain. Go into the refrigerated telescope room and see the largest optical/infrared telescope on earth. See if you can identify the hexagonal shapes of the mirror segments and determine which way the telescope is pointing. The Keck telescopes do not have mirror covers, so they are parked with the telescope pointing at the horizon, to minimize the possibility of dirt and moisture collecting on the mirror surface. There is a model sample of one of the hexagonal segments to your right as you face the telescope structure. After leaving the visitor's area, note that to the east (looking straight out from the Keck entrance), you can see all the observatories on the summit's ridge. This makes a great photo.

When you leave the Keck parking lot, you'll again be on a paved road; continuing east, you'll pass the silvery-domed NASA Infrared Telescope Facility on the left. In about a half-mile, you'll come to a stop sign, with the United Kingdom Infrared Telescope to your left and the University of Hawai'i 0.6-meter Telescope located about 100 yards to your right. A good place to watch the sunset is near the UH 0.6-meter Telescope.

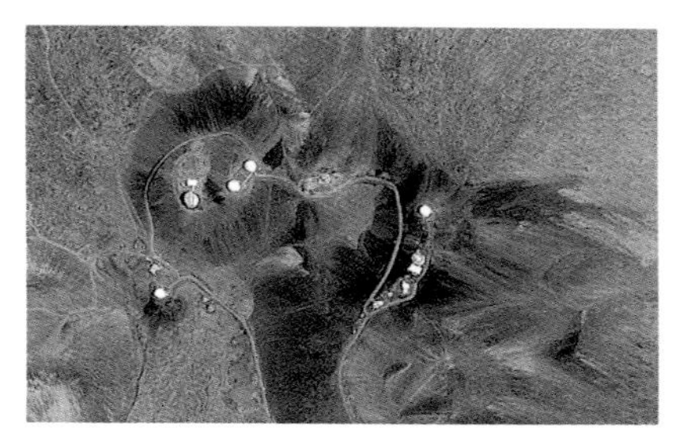

An aerial view of the summit shows the placement of the observatories.

Take a left turn at the stop sign. Passing the UKIRT on your left and continuing uphill, the next observatory you'll see is the UH 2.2-meter Telescope. Notice the crane, or horn, on the side of the dome. The observatory's Visitor Viewing Gallery is open 10 a.m. to 4 p.m. from Monday through Thursday, except state holidays. Be careful when walking up its four flights of stairs—this can be difficult at almost 14,000 feet (4,267 meters). If you feel faint or winded, just sit down on the stairs and rest for a while—even people who work at the observatories do it from time to time.

Driving by the UH 2.2-meter Telescope, the paving again stops as you pass the Frederick C. Gillett Gemini Telescope, or Gemini North. Notice the unique ventilation mechanisms on the mid-section of this great silvery dome. Continuing downhill from Gemini North, the last observatory is the beautiful Canada-France-Hawaii Telescope. Continue back to the UH 2.2-meter Telescope and park on the north or east side of the observatory.

Notice the Hawaiian *kuahu lele* (altar) on the nearby geographical summit to the east. In recent years some Hawaiians constructed it as an expression of their reverence for the summit and the mountain. You can show your respect for Hawaiian culture by not hiking to this sacred place. Anyway, the hike to the summit looks short, but it's strenuous because of the altitude and the loose cinders.

Mauna Kea's sunsets are some of the most beautiful in the world. In addition to the gorgeous colors and cloud formations, there is often a remarkable triangular-shaped shadow of Mauna Kea on the earth's atmosphere to the east as the sun sets in the west. These sunsets are one of the wonders of the Hawaiian Islands.

After sunset, you have a long, dark drive to the VIS ahead of you. You should be off the summit area half an hour after sunset. While among the observatories, try to minimize use of your headlights, and never use high beams. If you are able to, use only your parking lights and/or hazard flashers. Drive very slowly. Be sure to keep your vehicle in four-wheel drive, low range. Control the speed of your vehicle downhill by using a low gear. If you overuse your brakes, the thin air may cause them to overheat very quickly and cool slowly, causing brake failure.

Leaving the UH 2.2-meter Telescope after sunset, drive past the UKIRT and the UH 0.6-meter Telescope. Be careful on the steep, downhill section. There is a very sharp switchback about a quarter-mile downhill from the 0.6-m Telescope. Look toward the southeast as you approach this switchback—if there is enough light, you'll have a good view of the VLBA antenna.

In a short distance you'll arrive at the portable toilets where you began this part of the summit tour. Take a left turn, turn on your headlights and begin the half-hour drive back to the VIS. When you arrive there, you can join the free, public stargazing program.

This modern-day kuahu lele *is the site of traditional religious observances.*

ASTRONOMY IN HAWAIʻI

After all, the ancient Hawaiians were among the first great astronomers, using the stars to guide them among the islands in the vast Pacific, centuries before anyone else had developed such skill. Long before Europeans and mainlanders, Hawaiian astronomers were studying the heavens with awe and wonder, the same feelings that draw modern astronomers to study the heavens. At this very deep level, I feel we are brothers and sisters.

— Frederic Chaffee, Director,
W.M. Keck Observatory

Modern-day astronomers, such as those working at the observatories atop Mauna Kea, are certainly not the first people to use the Hawaiian skies for astronomical purposes.

The Hawaiian people, always closely attuned to their surroundings and environment, knew their night skies exceptionally well. Hawaiians distinguished between planets (*hōkū ʻaʻea* or *hōkū hele*) and fixed stars (*hōkū paʻa*). They were knowledgeable about stars moving across the sky from east to west, both nightly and throughout the year. Their measures of time and calendars based on these movements were impressively accurate. According to Hawaiian historian E. H. Bryan, more than a hundred Hawaiian star and planet names from ancient times are still known, and though they did not survive to modern times, likely more existed.

Some planets had names according to their position in the eastern or western sky, and some had fixed names, Bryan adds. Mercury, for instance, was called *Ukali*, or *Ukalialiʻi* (the names translates to "following the chief" [i.e., the sun] because it moved closely after the sun). Venus was known as *Mānalo*, as well as *Hōkū loa* ("morning star") when it was seen in the morning, and in the evening sky, *Nāholoholo* ("to flee, to run away"). What we call Jupiter the Hawaiians called *Kaʻawela* and *Ikaika* ("strong") because of its brightness.

Pre-contact Hawaiians knew and named star groups, as well. The Pleiades were known as *Makaliʻi* or *Huihui*. The appearance of *Makaliʻi* in the eastern evening sky at sunset after a new moon marked the beginning of the Hawaiian year and of a festival season still known as *Makahiki* (which we now also translate as "year"). This occurs around mid-November.

Together the belt and sword of Orion were known as *Nā Kao* ("the darts"). The Big Dipper is *Nā Hiku* ("the seven").

The North Star is *Hōkū-paʻa* ("fixed star") as well as *Hōkū-hoʻo-kelewa* ("steering or guiding star"). The Milky Way is *Kaʻu* ("something stretching across overhead"), *Iʻa* or *Hōkū-noho-aupuni*. Traditional voyagers heading to the south steered by a group of stars they called *Newa*, probably the Southern Cross.

Hawaiians call the sun *Lā*, which also means "day." The moon is *Mahina* (also the Hawaiian word for "month"), and Hawaiians specified a different name for each day of the month. *Hilo* was the day after the appearance of the new moon low in the west and *Hōkū*, the day after the full moon. There are 30 names for days in Hawaiian, but in some months only 29 were used.

NAVIGATION AND THE STARS

Traditional Hawaiian and Polynesian navigation could be the subject of its own book (and often has been—see the Reading List, pp. 137-39, for suggestions on further reading). Hawaiians inhabit these islands today because their ancestors were adventurous and talented wayfinders who left their Polynesian island homes and headed for unknown shores. We can only speculate why they left: perhaps they were driven out by warfare or maybe by famine; perhaps it was just a spirit of exploration. What we do know for sure, because some of their stories are told in ancient chants, is that they navigated the ocean not accidentally but purposefully. The ocean was a highway to Polynesians, who voyaged not with modern instrumentation but by watching natural signs such as currents, wind patterns, birds and the stars.

Another important marker to navigators was the sun. To stay on course, today's navigators align the position of the rising or setting sun with marks on the canoe's railing. Eight marks run down each side of the canoe, each one paired with a point at the canoe's stern that gives bearings in two directions. There are 32 bearings total, which match the 32 directional houses of the Hawaiian star compass.

Some people think ancient navigators observed the skies from atop Mauna Kea and that other celestial observations were made by using alignments with prominent features of the mountain.

In recent decades there has been a tremendous resurgence of interest in traditional Hawaiian navigation. Canoes have been crafted in the ancient style, navigators are again learning the old ways, and many successful voyages have been made between Polynesian island groups. And navigators-in-training are again studying the Hawaiian skies.

HISTORY OF MODERN ASTRONOMY ON MAUNA KEA

A 1960 tsunami, which did great damage to Hilo town, also adversely affected the local economy. Trying to revitalize it, the Hawaii Island Chamber of Commerce wrote to U.S. and Japanese universities with the idea of developing Mauna Kea and Mauna Loa as sites for astronomical observatories.

Dr. Gerard Kuiper, a famous planetary scientist from the University of Arizona, was already working with NASA and the Department of Defense to investigate Maui's Haleakalā for the same purpose, but he was hoping to find a location higher above the cloud layer. He discounted Mauna Loa because of the possibility of eruptions, but explored Mauna Kea's potential by persuading then-Governor John Burns to create a jeep trail up to the summit. Dr. Kuiper put a 12.5-inch (31-centimeter) telescope on Pu'u Poli'ahu to test the site, and the Arizona astronomer announced that "the mountaintop is probably the best site in the world—I repeat, in the world—from which to study the moon, the planets, and stars."

Nothing has changed: this is still the case. Mauna Kea's summit is usually cloudless and therefore has one of the highest numbers of clear nights per year in the world. Because the atmosphere is so stable at the mountain's peak, without disturbance from neighboring landforms, scientists can make more detailed observations from Mauna Kea than from elsewhere. And being above the tropical inversion cloud layer, the summit boasts skies that are pure, dry and without atmospheric pollutants. It's an astronomer's dream.

Construction of the NASA-funded University of Hawai'i 2.2-meter Telescope began in 1967. The University of Hawai'i Institute for Astronomy (IfA) was founded the same year. IfA astronomers teach astronomy at the undergraduate and graduate level, conduct astronomical research and manage astronomy development within the Mauna Kea Science Reserve. UH scientists receive guaranteed viewing time on all the telescopes on Mauna Kea.

In 1968, the Hawaii State Board of Land and Natural Resources granted a 65-year lease to the University of Hawai'i on the mountain's summit area, now known as the Mauna Kea Science Reserve. The observatories are near the middle of the summit plateau, and the remaining land serves as a buffer.

Two 0.6-meter telescopes were provided by the U.S. Air Force and NASA in the late 1960s, going into

Although there is no military presence on Mauna Kea today, the Department of Defense was one of the first entities interested in developing Mauna Kea.

operation in 1968 and 1969. The UH 2.2-meter Telescope followed in 1970 and was a tremendous success, producing images of unprecedented sharpness. Three major telescopes were then constructed in the 1970s, and the mountain summit was well on its way to becoming one of the top astronomy centers in the world.

On the following pages are overviews of each of Mauna Kea's observatories.

VERY LONG BASELINE ARRAY (VLBA)

The VLBA antenna on Mauna Kea is a radio antenna 25 meters (82 feet) in diameter. The VLBA is operated by the National Radio Astronomy Observatory (NRAO) in Socorro, New Mexico. The antenna is located at an altitude of 12,205 feet (3,720 meters), and it went into operation in 1993. The antenna surface is comprised of aluminum panels. The VLBA observes in 10 separate frequency bands ranging from 330 MHz to 86 GHz.

The VLBA is a group of ten 25-meter radio antennas that spans from St. Croix in the Virgin Islands in the east, across North America, to Mauna Kea in Hawai'i in the west. The data from each antenna are recorded on tape (soon it will be computer disks) and shipped to the Array Operations Center in Socorro, New Mexico, where they are correlated and the results sent to the various research scientists. Precise timing references are required to allow the various VLBA stations to operate as one instrument. This is accomplished by using a hydrogen maser that gives a timing accuracy of a few billionths of a second and a frequency stability of one part in a million billion.

Each antenna weighs 240 tons (217,724 kilograms) and is almost as tall as a 10-story building when the dish is stowed in an upright position. The VLBA antenna, installed on Mauna Kea in 1993, is the largest telescope on the mountain. The entire array, with baselines of up to 5,343 miles (8,599 kilometers) and frequencies up to 86 GHz, has the highest resolution of any dedicated telescope on Earth or in space. That resolution is less than one milliarcsecond (1/1000th of a second of arc). This is the equivalent of being able to read a small newspaper headline in Los Angeles from New York City. The

This radio image of the Crab Nebula was made with the Very Large Array radio telescope.

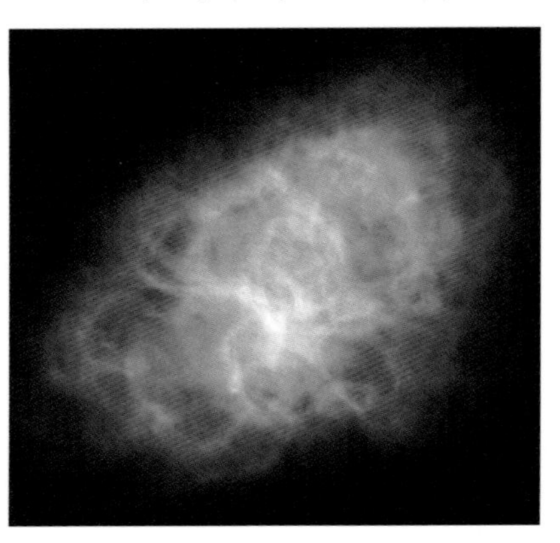

Did You Know?

The Very Long Baseline Array is a continent-spanning radio telescope that acts like a 5,000-mile (8,046-kilometer)-wide eye and produces the sharpest images of any earth- or space-based telescope.

Quick Info

First Light	1993
Aperture	25 m
Type	Radio
Website	http://www.vlba.nrao.edu

The VLBA antenna of Mauna Kea works with nine others as a single instrument.

VLBA occasionally participates in observations with other antennas around the world and in space to obtain even higher resolution.

The VLBA is used to study a wide range of astronomical objects. For example, a study of the galaxy NGC 4258, which is 21 million light-years distant, produced the most elegant and conclusive evidence for the existence of an extragalactic black hole. Other projects have used high-precision measurements of source positions to directly measure the rotation of our galaxy and to measure distances to, and motions of, pulsars. Many VLBA observations study the jets of energetic particles spewed out by black holes both in our galaxy and in the centers of other galaxies. On Earth, geophysicists can use VLBA data to learn about continental drift and the rotation and orientation of the Earth in space. Even for normal VLBA operations, the fact that Hawai'i is moving away from North America because of continental drift must be taken into account.

Mauna Kea's VLBA station is the only research telescope on the mountain not located at the summit area, nor is it visible from the Mauna Kea Access Road. Take a few minutes to drive down the short road and view this largest of all Mauna Kea telescopes, and try to imagine an eye that is 5,000 miles (8,046 kilometers) wide.

Note that this telescope is very sensitive to radio interference. Do not use cell phones anywhere near it.

CALTECH SUBMILLIMETER OBSERVATORY (CSO)

The Caltech Submillimeter Observatory (CSO) houses a radio telescope that operates in the region of the electromagnetic spectrum just beyond the infrared region in wavelength, hence the name submillimeter. It is operated by the California Institute of Technology with headquarters in Pasadena, California. The telescope is located near the summit of Mauna Kea at an altitude of about 13,375 feet (4,077 meters). Commissioned in 1987, it remains one of the world's premier submillimeter telescopes.

The CSO's collecting dish is 10.4 meters (34 feet) in diameter and is made up of 84 hexagonal aluminum panels, individually adjustable for optimum surface figure. The dish and its back-up structure were designed by the late Dr. Robert B. Leighton, a physics professor at Caltech. The dish is mounted so that it can point to any part of the sky. Since it is housed in a very compact dome, the entire building must rotate when the dish is rotated. During the daytime and in inclement weather a shutter closes to protect the telescope. In principle the telescope could be operated during the daytime, but the heat from exposure to the sun would cause thermal expansion, which would distort the shape of the dish.

Radiation collected by the dish is fed by mirrors to sensitive detectors that must operate at temperatures of liquid helium, making use of superconducting materials. The receivers are capable of responding to radiation at frequencies from about 180 to 850 GHz (about 1.7 mm to 0.35 mm in wavelength).

Emissions in this millimeter to submillimeter range are generated

Did You Know?

The Caltech Submillimeter Observatory is the only professional observatory in the world without a specific operator. The observing astronomer has full control of the telescope during his or her observational period.

Quick Info

First Light	1987
Aperture	10.4 m
Type	Submillimeter
Website	http://www.submm.caltech.edu/cso

CSO observed the submillimeter emissions superimosed as contours on this picture of the Helix Nebula.

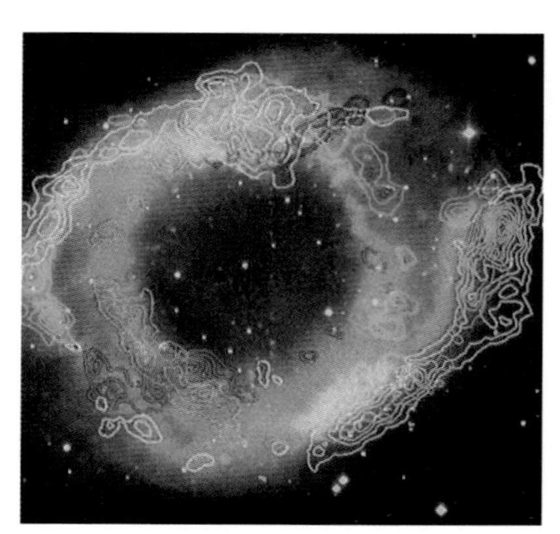

CSO's futuristic exterior matches the high-tech equipment within.

by cool dust and gases in many types of objects in the universe, including molecular clouds like the great Orion Nebula (M42), planetary nebulae like the Ring Nebula (M57), and external galaxies like the Whirlpool Galaxy (M51). These objects can also be seen with optical telescopes at the Visitor Information Station's stargazing program. However, the submillimeter views are quite different from the optical views!

Dr. Leighton, the principal architect of this telescope, was also the team leader at the Jet Propulsion Laboratory for the Mars probes Mariner 4, 6, and 7 and helped to develop a digital camera system for use in deep space. Leighton is also known for discovering the five-minute oscillations in the local surface velocities of the sun. This discovery began the field known as solar seismology.

THE JAMES CLERK MAXWELL TELESCOPE (JCMT)

The JCMT is a submillimeter telescope with an antenna 15 meters (49 feet) in diameter. The JCMT is operated by the Joint Astronomy Centre in Hilo and funded by the United Kingdom, Canada and the Netherlands. It is located at an altitude of 13,428 feet (4,093 meters), and first light was achieved in 1987. The antenna surface is constructed of 276 aluminum panels. The JCMT typically observes at frequencies between 150 GHz and 870GHz.

James Clerk Maxwell (1831-1879), a Scottish mathematician and physicist, developed and published the unified formulation of electromagnetic waves. We know these equations today as the Maxwell Equations. Maxwell's first major contribution to science was the study of Saturn's rings, which he argued were made of small individual solid particles. The first *Voyager* spacecraft to reach Saturn confirmed his theory. A small gap in Saturn's rings, the "Maxwell Gap," is named after him. Maxwell is also responsible in part for the first color photograph, having developed the "trichromatic process" that is the basis of modern color photography.

The JCMT is the world's largest single-dish submillimeter telescope. It's as wide as a basketball court. The entire dish and related structure are enclosed in a carousel for protection from the weather. When observing, the roof and door are open, revealing the largest piece of Gore-Tex in the world, which is attached to the opening in the enclosure to protect the telescope from wind, dust and moisture. This also allows the telescope to be pointed safely at or near the sun, in order to observe the sun itself or the inner planets. The Gore-Tex is transparent to submillimeter radiation.

The JCMT is specifically designed to observe radiation in the "submillimeter" region of the electromagnetic spectrum, a region between infrared radiation and radio waves. This type of radition does not come from stars but from cooler material in the universe, such as clouds of gas or dust found between stars. This material, commonly called the interstellar medium, is the mate-

Did You Know?

The JCMT is the world's largest single-dish submillimeter telescope. One of the coldest continuously chilled places on earth, and perhaps in the universe, is located in one of the detectors at the telescope. It's kept at less than one tenth of a degree above absolute zero, or –459.9 degrees Fahrenheit (-273.3 degrees Celsius).

Quick Info

First Light	1987
Aperture	15 m
Type	Submillimeter
Website	http://jach.hawaii.edu/outreach (shared with UKIRT)

rial from which new stars form as well as what is left over when they die. Water vapor in the atmosphere usually absorbs this type of light before it can be observed, but on Mauna Kea much of the earth's atmosphere is below the summit, which, coupled with the dry air at that altitude, makes it an ideal location for this type of telescope.

To make an image at these wavelengths, a revolutionary type of instrument has been developed. This device is called the Submillimetre Common User Bolometer Array, or SCUBA for short. This is like a regular camera, except it can "see" in the submillimeter realm. To achieve this, the detectors must be kept very cold. In fact, one of the coldest continuously chilled places on earth is located inside this instrument. It is less than one tenth of a degree above absolute zero, or -459.5 degrees Fahrenheit (-273.3 degrees Celsius). A second-generation submillimeter camera called SCUBA-2 will replace SCUBA and will be more than 1,000 times more sensitive for producing images of large areas of the sky.

You can find your "birthday star" by going to the Joint Astronomy Centre Outreach web page (*http://outreach.jach.hawaii.edu*) and clicking on "birthday stars." By typing in your birthday, you'll find a star that is approximately your age in light years away. This means the light we see from that star today actually left the star about the time you were born—it has taken your entire lifetime to reach your eyes.

The biggest piece of Gore-Tex in the world (it's transparent to the light that this telescope studies) is attached in front of the telescope.

SUBMILLIMETER ARRAY (SMA)

The SMA, located at an altitude of 13,386 feet (4,080 meters) is a collaborative project of the Smithsonian Astrophysical Observatory in Cambridge, Massachusetts, and the Academia Sinica Institute of Astronomy & Astrophysics of Taiwan. First light was achieved in 2002. The antenna surfaces are constructed of high-precision machined aluminum panels attached to a very stiff carbon-fiber support structure. The SMA observes in the "submillimeter" band, which refers to the wavelength of light the array detects—in this case, 0.3 to 1.7 millimeters (0.01 to 0.07 inch or 170 to 920 GHz). It is the world's first interferometer dedicated solely to submillimeter astronomy.

Submillimeter emission is readily absorbed by water vapor in the earth's atmosphere. By the time this radiation reaches sea level, the signals are so faint that it is impossible to make astronomical observations. Because of this, submillimeter observatories must be built on high, dry locations. Mauna Kea is one of the best sites in the world for this type of observatory.

The SMA consists of eight portable antennas, each with a diameter of 6 meters (19.7 feet) and each weighing 94,600 pounds (42,910 kilograms). These antennas are always used together and cannot make observations independently. Thus the SMA is one telescope with eight "mirror" elements. This results in 28 simultaneous baselines (the "3-D distance" from one antenna to another). There are 24 different antenna-mounting pads configured in four rings, which produce baselines from 26 to 1,666 feet (8 to 508 meters—the length of five football fields). The antennas are moved between these pads by a special

Did You Know?

The Submillimeter Array on Mauna Kea is the world's first interferometer dedicated to submillimeter astronomy.

Quick Info

First Light	2002
Aperture	8 dishes, each 6 m in diameter
Type	Submillimeter
Website	http://sma-www.harvard.edu

Here is an image of M51 made with the SMA showing the location and movement of CO gas in the galaxy.

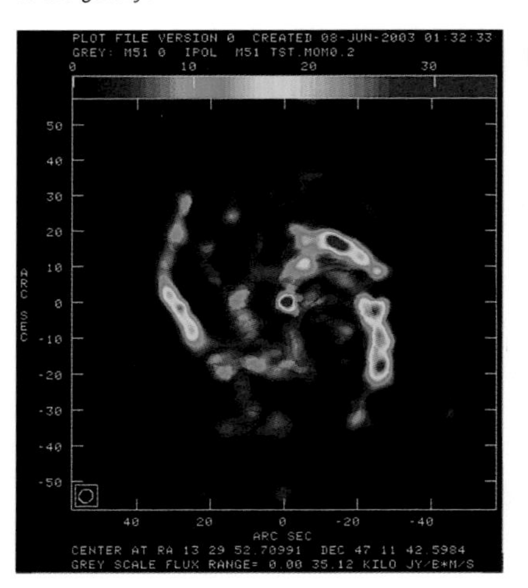

The eight telescopes of the Submillimeter Array work as a single unit.

transporter. This ability to move the antennas to different locations allows SMA researchers to have many hundreds of different baselines, which makes the array very flexible and able to carry out many different kinds of observations.

Data from each of the antennas are transmitted through fiber-optic cables to the control building located adjacent to the array. There the signals are compared to each other with a special supercomputer called the "Correlator." The highest resolution of the SMA is 0.1 arcseconds. If your eyes had such high resolution, you'd be able to see the period at the end of a sentence from a distance of one mile! Starting in 2005, the other two submillimeter telescopes in Submillimeter Valley, CSO and JCMT, will occasionally be incorporated into the Submillimeter Array. This will further increase the sensitivity and resolving power of the combined observatories by increasing the number of baselines to 45, doubling the point-source sensitivity and increasing the resolution by about 50 percent.

When viewing the array, try to discern the circular ring placement of the antennas. Be careful to stay well away from the antennas, though, because they are remotely operated and may move very quickly at any time. The speed limit on this unpaved portion of the road is 5 mph because dust can damage these antennas and receivers.

SUBARU TELESCOPE

Subaru Telescope is an optical infrared telescope with a mirror diameter of 8.3 meters (27 feet). The National Astronomical Observatory of Japan, part of the National Institutes of Natural Sciences, operates the observatory. It is located at an altitude of 13,658 feet (4,163 meters), and first light was achieved in 1999. The mirror is made of ultra-low thermal expansion glass and is 20 centimeters (8 inches) thick. Subaru observes over the full range of wavelengths from ultraviolet to mid-infrared.

Subaru Telescope is Japan's premier telescope and its most well known national research facility. *Subaru* is the Japanese name for the open star cluster we call the Pleiades, also known in Hawaiian as *Makali'i* and in English as the Seven Sisters. Construction of the telescope began in 1991. The entire telescope weighs 555 tons (504,000 kilograms) and is 73 feet (22.2 meters) high.

This image of the Dwarf Irregular Galaxy Leo A was taken by the Subaru Telescope.

Did You Know?

The 8.3-meter primary mirror of the Subaru Telescope is the largest and smoothest single-piece mirror on Mauna Kea.

Quick Info

First Light	1999
Aperture	8.3 m
Type	Optical Infrared
Website	http://www.subarutelescope.org

Subaru's primary mirror is one of the largest monolithic mirrors in existence and is the largest that is in regular use and available to all astronomers in the world. Sometimes the telescope is referred to as having an effective aperture of 8.2 meters (26.9 feet). That is because only the light falling on the inner part of the mirror is used. It took three years to manufacture the mirror blank and another four years to polish the mirror and drill holes in the back of the mirror for the 261 actuator fingers that provide the "active support system," which keeps the mirror in a precise shape no matter where it is pointing.

Instead of a dome with the common hemispherical design, Subaru has an aluminum enclosure with a cylindrical shape. Computer simulations and hydrodynamic experiments showed that this was the best shape to prevent warm and turbulent outside air from entering the enclosure. To reduce heat sources within the enclosure, the telescope control electronics and observing room are in a separate building. The cylindrical enclosure rotates with the telescope.

Subaru has seven first-generation instruments that can capture images and spectra at different wavelengths of light from the visible to the infrared. Several robots help attach these instruments and the secondary mirrors to the telescope when needed.

Subaru is the only telescope with a primary mirror 8 meters (26 feet) or larger in diameter that has an instrument at prime focus (at the top of the telescope, above the mirror). Telescopes usually use a secondary mirror mounted at the top of the telescope to reflect light to a focus near the bottom of the telescope. Instruments at prime focus have a larger field of view than instruments at other foci. This makes Subaru one of the most efficient telescopes for searching for large numbers of very faint astronomical objects at the edge of our universe or of our solar system.

Free tours of Subaru are offered two or three days a week at 10:30 a.m., 11:30 a.m. and 1:30 p.m., in English and Japanese. Individuals and groups may sign up for tours on a first come, first served basis by visiting the Subaru Telescope website. For more information, contact the Subaru Summit Tour Inquiry Hotline at 934-5056 or visit the web page previously mentioned.

Subaru Telescope can be seen from almost any vantage point on the summit area. It is easy to identify because of its unique cylindrical enclosure.

Subaru's cylindrical enclosure was chosen to minimize air turbulence.

W. M. KECK OBSERVATORY (KECK I AND KECK II)

The W. M. Keck Observatory, located at an altitude of 13,647 feet (4,160 meters) houses two optical infrared telescopes, the mirrors of which are each 10 meters (33 feet) in diameter. California Institute of Technology, the University of California and the National Aeronautics and Space Administration operate the observatory. Each mirror is made of 36 hexagonal segments. Each hexagonal segment is 1.8 meters (6 feet) wide and 7.5 centimeters (3 inches) thick and is made of Zerodur™, a low-expansion glass-ceramic. The telescopes observe in the visible band and the infrared.

In its first decade of operation, the W. M. Keck Observatory established itself as a world leader in astronomical research. Scientific breakthroughs that emerged from the Keck Observatory in its early years forever changed the world's understanding of the cosmos and made the Observatory an icon of astronomical discovery.

Today the Keck telescopes are used to seek answers to universal questions such as, How did the universe evolve? How and when did galaxies first form? What is dark matter or its more mysterious counterpart, dark energy? What do black holes teach us about high-energy environments? Are there planets outside our solar system that show chemical signatures of life? What is the ultimate fate of the universe?

Did You Know?

The Keck I telescope has confirmed the discovery of more extra-solar planets than any other ground-based telesccope on Earth.

Quick Info

First Light	Keck I, 1992, Keck II, 1996
Aperture	10 m x 2
Type	Optical Infrared
Website	http://www2.keck.hawaii.edu

Visitor's Area—Open 10 a.m. to 4 p.m., Monday through Friday

The twin Keck telescopes have a popular Visitor's Area and an informative website.

To answer these and other questions, teams of engineers, technicians and scientists use faster computers and new optical and detector technology to drive astronomical science into a new era. For example, the twin Keck telescopes can now be used together as a single instrument to obtain the effective resolution of an 85-meter (279-foot) telescope when looking in the near-infrared. New optical technology, including an artificial laser guide star, has allowed scientists to virtually remove all effects of atmospheric blurring from Keck images, providing astronomical images from the ground that are better than those taken from space. Wide-field imaging and new spectral instruments are allowing astronomers to extract more information from each exposure than ever before—obtaining hundreds and sometimes thousands of spectra in a single exposure. In the next few years these advances will take the W. M. Keck Observatory into the realm of omni-wavelength, diffraction-limited performance—the point at which each Keck telescope performs to its theoretically achievable design limit.

Several new technologies were developed to build the Keck mirrors, each of which is made up of 36 hexagonal segments. "Stressed mirror polishing" was specifically developed to shape these segments. Each circular mirror blank was placed in a special vise and deformed in a precise pre-determined manner, and a simple spherical shape was polished into the surface. The segment was then removed from the vise, and the mirror "relaxed" into its desired asymmetrical shape. The surface of each segment is so smooth that if it were as wide as the earth, its highest mountain would be only three feet high!

When assembled into a complete mirror, each mirror segment is stabilized by a system of extremely stable support structures. A computer-controlled system of sensors and actuators (or precision pistons) adjusts the position of each of these segments twice per second. This results in an accuracy of 4 nanometers (about 1,000 times thinner than a human hair) for each segment.

The late William Myron Keck, founder of The Superior Oil Company, established the W. M. Keck Foundation in 1954. Following Mr. Keck's philosophies of imagination, innovation and new technology, the W. M. Keck Foundation is committed to using imagination and innovation in giving grants to support new scientific discoveries and technologies.

You can't miss seeing the twin Kecks on the west side of the summit. Be sure to visit the Visitor's Area, open Monday through Friday from 10 a.m. to 4 p.m. It is located in the building to the south. In addition to the interpretative panels, there are also two nice bathrooms and a drinking fountain. You can actually go into the telescope viewing room and see partial views of the Keck I telescope and the dome. The dome is cooled to nighttime temperatures, so dress warmly.

NASA INFRARED TELESCOPE FACILITY (NASA IRTF)

The NASA Infrared Telescope Facility is an infrared telescope with a mirror 3 meters (10 feet) in diameter. The observatory is operated and managed by the University of Hawai'i Institute for Astronomy in Honolulu and Hilo. The telescope is located near the summit of Mauna Kea, at an altitude of 13,675 feet (4,168 meters) and was commissioned in 1979. The mirror is made of Cervit 201™ and is 13.7 inches (34.5 centimeters) thick at the edge. The NASA IRTF is optimized for observing in the infrared, at wavelengths between 1 and 25 microns.

The primary mission of the IRTF is to provide observations in support of NASA's space missions and research programs. Half of the observing time on the IRTF is allocated to observing objects in our solar system. Astronomers are granted time to monitor many different bodies orbiting our sun, including asteroids, comets, and the planets and their satellites, so they can learn about the properties of these objects and how they change over time.

One exciting program that the IRTF has been involved in for more than 15 years is monitoring the volcanic activity on Io, one of the large moons of Jupiter. Io is the most geologically active body in the solar system, with dozens of active volcanoes continually changing its surface. In the infrared, these volcanoes appear as "hot spots" that can be observed using telescopes on the Earth. One of Io's volcanoes, called Loki, has been studied particularly closely, and it has been observed to have recurring periods of activity. Loki's periodic behavior has been interpreted as possibly being due to a lava lake that heats up from time to time and overturns. This behavior might be similar to that of the lava lake that was in Kīlauea Volcano's Halema'uma'u Crater, except that for scale, Loki's lava lake would be the size of the entire Big Island.

At the IRTF, a variety of instruments are used to learn about objects in our solar system.

Did You Know?

The NASA IRTF is the only ground-based telescope dedicated to the support of spacecraft missions and basic solar system research.

Quick Info

First Light	1979
Aperture	3 m
Type	Infrared
Website	http://irtfweb.ifa.hawaii.edu

The NASA IRTF is one of three telescopes on Mauna Kea that became operational in 1979.

Other research projects being conducted at the IRTF include monitoring Saturn's atmosphere, rings and satellites in support of NASA's Cassini mission; mapping water and methane in the atmosphere of Mars; and measuring the compositions of asteroids and comets. The IRTF is also being used to observe very young stars in other parts of our galaxy, and is helping us learn how planets form around other stars.

When you drive by this telescope, notice its silver dome and the square shape of the first level. This unique design makes it easy to identify.

UH 0.6-METER TELESCOPE

The UH 0.6-meter Telescope is an optical/infrared telescope with a mirror diameter of 0.6 meters (24 inches). The University of Hawai'i operates the observatory. It is located at an altitude of 13,735 feet (4,186 meters), and first light was achieved in 1968. The telescope was made by Boller and Chivens, and the primary mirror material is 4.75-inch (12-centimeter)-thick Pyrex. Observations are made in the visible band and the near-infrared.

The UH 0.6-meter Telescope was built by the U.S. Air Force in 1968 and was operational for the first lunar landing in June 1969. It is the oldest operational telescope on Mauna Kea. In its early days, planetary research was done using photographic film and early infrared detectors. It was the first telescope to use one of the NICMOS infrared detector arrays, which were developed for the Hubble Space Telescope.

This image of the Horsehead Nebula was captured by the UH 0.6-meter Telescope.

Did You Know?

The University of Hawai'i 0.6-meter Telescope is the oldest telescope in operation on Mauna Kea. It is also the smallest on the summit area.

Quick Info

First Light	1968
Aperture	0.6 m
Type	Optical
Website	http://www.ifa.hawaii.edu

The telescope was used with several instruments in the 1990s, including CCD cameras, infrared cameras and photometers, as well as a Danish Star Tracker instrument. Since 2001, the telescope has been used primarily by the faculty and students of the UH Hilo Department of Physics and Astronomy, thanks partly to funding from the NASA New Opportunities through Minority Initiatives in Space Science Program.

The telescope is also used for research projects involving collaborations between the University of Hawai'i at Hilo faculty and investigators at other institutions, such as Iowa State University, University of Vienna, Ohio State University and Rochester Institute of Technology. Some of the scientific goals of this research have included high-speed photometry (brightness measurements) of white dwarf stars; photometry of hot blue and cool yellow variable stars; CCD imaging of stars in

This small telescope continues to teach new generations of astronomers.

young clusters; spectrophotometry (brightness measurements in different colors) of young forming stars in the Orion nebula; CCD photometry of the center of the Milky Way Galaxy; and searches for planets or brown dwarfs around nearby stars with microlensing.

The UH 0.6-meter Telescope is the only telescope on Mauna Kea that is operated by astronomers exposed to the cold night air. The dome has no heated control room. The frigid nighttime temperatures at the summit combined with wind and the reduced oxygen available at this altitude can make observing at this telescope something of a challenge.

There is a small area for parking near the dome. This is a popular place to view the sunset. Portable chemical public toilets are located nearby.

UNITED KINGDOM INFRA-RED TELESCOPE (UKIRT)

UKIRT is an infrared telescope with a mirror diameter of 3.8 meters (12.5 feet). The Joint Astronomy Centre at Mānoa operates the telescope and the UK's Particle and Astronomy Research Council. It is located at an altitude of 13,775 feet (4,198 meters), and first light was achieved in 1979. The mirror is made of Owens-Illinois Cervit™ and is 287 millimeters (11 inches) thick at the outer edge. The telescope observes at wavelengths between 1 and 30 microns.

UKIRT started out as a low-cost facility with modest optical performance requirements, though it was built with an extremely high-quality primary mirror. Over the years, many improvements have been made, and now the best UKIRT images rival those of the Hubble Space Telescope's NICMOS infrared imager in sharpness. This is a good example of how an "older and smaller" telescope can continue to be competitive through improved technology.

"Infrared" comes from two words: "infra," meaning "below," and "red," referring to the red component of light. It is fitting that the United Kingdom operates the largest telescope dedicated to infrared astronomy, as it was the birthplace of this science. William Herschel, who was born in Hanover, Germany, in 1738 and moved to England in 1755, is best known for his 1781 discovery of the planet Uranus, the first planet to be discovered since antiquity. This discovery greatly expanded our understanding of our solar system. Less known, but of greater significance to astronomy and physics, is Herschel's 1800 discovery of infrared radiation. He was measuring differing levels of heat from different colors of sunlight as they passed through a prism. To his surprise, he found that the greatest amount of heat

This image of Orion was created by UKIRT.

Did You Know?

UKIRT is the world's largest telescope dedicated solely to infrared astronomy.

Quick Info

First Light	1979
Aperture	3.8 m
Type	Infrared
Website	http://outreach.jach.hawaii.edu (shared with JCMT)

was beyond the red part of the spectrum, where no color could be seen by the human eye. With this groundbreaking observation, infrared astronomy was born.

Fluctuations in temperature can cause several adverse effects in infrared observations, but efforts are made to minimize them. "Facility-seeing" effects are caused by differences in the temperature of the air inside the dome and that of the air outside. This mixing of warmer and colder air results in a blurring of the images of astronomical objects viewed through this mixed air. To minimize the effect, UKIRT uses forced ventilation (an extractor fan) and natural ventilation (wind blowing through a set of 16 controllable openings in the dome). "Mirror seeing" is caused when the mirror is warmer than the air around it. This causes a small amount of turbulence in the air above the mirror and reduces the quality of the images. During the daytime, cold air is blown across the mirror to keep the mirror temperature at the colder, forecast nighttime temperature.

When you drive by UKIRT, notice the square, controllable openings in the dome. You'll be able to see easily how the cold night air can pass through the small compact dome.

Ventilation ports around the center of the UKIRT dome help keep the temperature inside equal to the outside temperature.

UH 2.2-METER TELESCOPE

The UH 2.2-meter Telescope is an optical infrared telescope with a mirror diameter of 2.2 meters (88 inches). The Institute for Astronomy, University of Hawai'i at Mānoa, operates the telescope. It is located at an altitude of 13,824 feet (4,213 meters), and first light was achieved in 1970. The mirror is made of fused silica and is 291 millimeters (11.5 inches) thick. The telescope observes in the ultraviolet, the visible band and the near-infrared.

NASA funded the University of Hawai'i 2.2-meter Telescope. It was the first large-scale telescope built on Mauna Kea and the first to be built at high altitude. The telescope was also one of the first to have computerized control systems. It demonstrated the virtues of Mauna Kea to the worldwide astronomical community and led the astronomical development of the summit. When viewing the observatory's dome from the outside, note the large crane located at the highest point in the dome. This "horn," the distinctive feature of the observatory, is actually a 20-ton crane used for moving heavy pieces of equipment, such as those used to aluminize the mirror, from outside into the dome and to any location on the observing floor. It is also used for moving equipment onto the Cassegrain observing platform.

In 1992, the UH 2.2-meter Telescope was used to discover the first object in the Kuiper Belt—a region of the solar system containing vast numbers of small bodies outside the orbit of Neptune. Much of what we know about the Kuiper Belt came from observations with the UH 2.2-meter Telescope in the early 1990s. Research on the Kuiper Belt continues to be a hot topic in astronomy; many of the larger telescopes on Mauna Kea are being used to study these very faint objects that are

Did You Know?

The University of Hawai'i 2.2-meter Telescope has a mirror about the same size as the Hubble Space Telescope.

Quick Info

First Light	1970
Aperture	2.24 m (88 in)
Type	Optical Infrared
Website	http://www.ifa.hawaii.edu/88inch/

Visitor Viewing Gallery—Open 10 a.m. to 4 p.m., Monday through Thursday (except state holidays)

This infrared image of the star formation region M17 was obtained with the large-format infrared camera at the UH 2.2-meter Telescope.

Check out this observatory's Visitor's Gallery.

relics from the formation of our solar system. Most astronomers now believe that Pluto is not a planet, but is simply the largest known Kuiper Belt Object.

Since 2000, UH astronomers have discovered or confirmed 60 new moons of the giant planets from telescopes on Mauna Kea, including the UH-2.2-meter Telescope, Subaru and Keck. At the time this book was printed, there were 63 known satellites of Jupiter, 33 of Saturn, 27 of Uranus and 46 of Neptune.

Other solar system objects studied with this telescope include Trojan asteroids, comets and asteroids. The telescope has also been a test bed for new technology, including the revolutionary Orthogonal Transfer Charge Coupled Device (OTCCD) and the Ultra Low Background nfrared camera, which has over 16 million pixels and was the first very large infrared camera in the world.

Be sure to go into the Visitor's Gallery, which is open Monday through Thursday from 10 a.m. to 4 p.m., except state holidays. Be careful when walking up the several flights of stairs—remember there is 40% less oxygen available for you to breathe. If you start to feel faint, just sit down on the steps and rest until you feel better.

In the Visitor's Gallery, the entire telescope room is visible through a glass window. There are also several interpretative panels. Note the instrumentation cluster on the bottom of the telescope.

There is parking on all sides of this observatory, but be sure to park well away from the building if there is snow and ice on it, to avoid injury from falling ice. There's a good view of the sunset from the west side. Be sure to turn around during sunset and see Mauna Kea's shadow to the east. It's usually well-defined and makes a spectacular photo.

THE FREDERICK C. GILLETT GEMINI TELESCOPE (GEMINI NORTH)

Gemini North is one of a pair of telescopes with mirror diameters of 8.1 meters (26.5 feet). The U.S. National Science Foundation (NSF) funds almost half of the Gemini Observatory and acts as the executive agency for its seven international partners, the United States, the United Kingdom, Canada, Australia, Brazil, Argentina and Chile. Gemini North is located at an altitude of 13,824 feet (4,213 meters), and first light was achieved in 1999. The mirror is made of ultra-low-expansion (ULE) glass from Corning, and it is about 20 centimeters (8 inches) thick. The telescope observes in the visible band and the infrared.

The Gemini Observatory is composed of twin telescopes, one located on Mauna Kea (Gemini North) and the other on north-central Chile's Cerro Pachón (Gemini South). With a telescope on each hemisphere, the observatory can observe objects in the entire night sky.

The Gemini Observatory uses some of the most advanced technologies available, exemplified by its mirror-coating facilities. Each of the twin telescopes has a facility that can coat the 8.1-meter mirror with either aluminum or protected silver using what is called "sputtering" technology. It takes about 16 grams (half an ounce) of aluminum, a little more than you would find in a soda can, to coat the primary mirror.

Did You Know?

Though it weighs approximately 380 tons (345,000 kilograms), the Gemini North telescope structure rides on hydrostatic bearings, and when the drive motors are disengaged and its inertia is overcome a single person could move it.

Quick Info

First Light	1999 (Gemini North)
Aperture	8.1 m
Type	Optical/Infrared
Website	http://www.gemini.edu

The silver coatings produce unparalleled data from infrared sources, such as where stars (and possibly planets) are performing. Gemini is the only observatory with the technology to coat 8-meter-class mirrors with a protected silver coating, first accomplished in 2004 and requiring about two ounces of silver. Using a protected silver coating rather than aluminum, the telescope is able to obtain data that is equivalent to increasing the mirror size to more than 11 meters (36 feet) with some types of infrared light. This technological advance places the Gemini Observatory in the forefront of infrared astronomy.

One of the most beautiful sights in the night sky is the twinkling stars. Starlight "twinkles" because it is distorted by turbulence in the earth's atmosphere, and though it looks beautiful, it causes problems for astronomers. One of the reasons the Hubble Space Telescope was put into space was to eliminate the blurring effects of the atmosphere.

To reduce this blurring effect, Gemini and most large modern observatories have utilized a technology called adaptive optics (AO). AO works by using a device called a wavefront sensor to rapidly sample the starlight and determine how the atmosphere has distorted it at any given moment. This information is processed and sent to a small deformable mirror (about the size of an adult's palm), which can change its shape up to 1,000 times per second to cancel out most of the distortions to the starlight. This produces remarkably high-resolution images. For Gemini the clarity with AO is equivalent to seeing the separation between a car's headlights 2,500 miles away.

One of the constraints when using AO systems is that a "guide star" must be available to sample the incoming light. This is a problem because the guide star must be close to the object being studied; the farther away the guide star, the less effective the correction is. To address this problem, Gemini Observatory has developed a solid-state laser to create an artificial laser guide star. This has been incorporated into Gemini's current AO system and will be used for a new system called Multi-Conjugate Adaptive Optics, or MCAO, that uses multiple laser guide stars. This system also uses multiple deformable mirrors that are turned to specific layers of the earth's atmosphere, and allows for corrections of an area of the night sky over 25 times larger than possible with current AO technology.

Notice Gemini's large vents, which are opened during observations. These allow the cold night air to flow through the observatory, which produces more stable images. The silver-colored dome produces better thermal stability than the more traditional white domes.

A Gemini Virtual Tour is available at the VIS's computers. Be sure to try it—it's as close to being there as you can get without the effects of altitude!

Gemini North is shown ready to observe the night sky.

CANADA-FRANCE-HAWAII TELESCOPE

The Canada-France-Hawaii Telescope (CFHT) is an optical infrared telescope with a mirror diameter of 3.6 meters (12 feet). Canada, France and the University of Hawai'i, through a tripartite agreement, operate the telescope jointly from the ranching town of Waimea on the island of Hawai'i. It is located at an altitude of 13,793 feet (4,204 meters), and first light was achieved in 1979. The mirror is made of the low-expansion coefficient glass-ceramic Cervit™ and is 0.6 meters (2 feet) thick. The telescope observes in the visible band and the infrared.

The CFHT was the first observatory on Mauna Kea to have several international partners. The architectural shape of its enclosure is considered by many to be the most beautiful on the summit. Notice how it seems to float in the air.

When it was built in 1979, the 3.6-meter telescope was the sixth largest in the world. Telescopes built since the 1990s, however, typically have mirror sizes in the range of 8 to 10 meters (26 to 33 feet). To compensate for the disadvantage of its "small" mirror size, CFHT has begun an aggressive instrumentation development program to come up with leading-edge technology in order to remain competitive with the new generation of larger telescopes.

Most 8- to 10-meter telescopes are designed to collect large quantities of light in order to observe faint objects, but they typically have a small field-of-view (i.e., they can see only a very small part of the sky). An advantage of many smaller telescopes is that they have a wider field-of-view, and so can see a larger area of the sky.

To take advantage of this, CFHT developed the MegaPrime imager with the MegaCam camera. This wide-field imager is composed of 36 2048 x 4612 pixel CCDs (or a total of 340 megapixels). This can cover almost a full 1 degree x 1 degree field-of-view with a resolution of 0.187 arcseconds per pixel. Each image from this camera is about 700 megabytes, enough to fill one complact disc with data. This ensures proper sampling of the sharp images achieved on Mauna Kea. You could say the MegaCam is the biggest camera in the world; it can image an area of the sky larger than four full moons.

This camera began operating in January of 2003 and has gathered

Did You Know?

The MegaPrime imager is the largest closed-packed array in the world (18,448 x 18,432 pixels).

Quick Info

First Light	1979
Aperture	3.6 m
Type	Optical Infrared
Website	http://www.cfht.hawaii.edu

Time-lapse photography shows "star trails," the paths the stars appear to move in, above the CFHT.

an amazing amount of data, placing CFHT on the leading edge of the scientific competition. A wide field-of-view infrared camera is currently under development. This will complement the optical MegaCam and allow wide field-of-view imaging in the infrared. It is scheduled to become operational in 2005.

The routine use of adaptive optics (AO, see p. 117) pioneered at CFHT and led to the first image from a ground-based telescope of a moon orbiting an asteroid. "Petit-Prince" is the name given to this little moon by the team of astronomers who discovered it around Eugenia, an asteroid of the main-belt. Without AO, Petit-Prince would have been lost in the glare of Eugenia. More moons of asteroids and binary asteroids have been discovered at CFHT and later at other observatories since AO has become available on 8-m-class telescopes, each time leading to the determination of the mass and the density of the primary asteroid, both key parameters for the understanding of the nature of these small bodies of our solar system.

CFHT produces a high-quality calendar annually. A different set of astronomical images, some of the most beautiful objects in the night sky, all taken from the observatory, is featured every year. A set of full-size posters is also available; both can be purchased at the VIS's First Light Bookstore. ▲

7 Maunakea Discovery Center

7 Maunakea Discovery Center

Titanium cones represent the Big Island's highest volcanoes.

Opening late 2005, the Maunakea Discovery Center (MDC) is located on the corner of 'Imiloa Place and Nowelo Street in Hilo. Take Waiānuenue Avenue up from the bay and turn left on Komohana Avenue. Nowelo Street will be on your left. The center sits in UH Hilo's Science and Technology Park. Several of the Mauna Kea observatories maintain base facilities at this research and technology park.

If you are in Hilo before heading up to Mauna Kea, plan to visit the Maunakea Discovery Center first to make your experience on the mountain a richer one. Going through the center in the morning, for instance, and then driving up to the mountain in the afternoon makes for a great itinerary and a full Mauna Kea experience.

The brand new, $28-million "edu-tainment" center, funded by NASA, illustrates and integrates two very different aspects of Mauna Kea—high-tech science and traditional Hawaiian culture. Through its many interactive exhibits, planetarium shows, educational programs and special events, visitors are immersed in the astronomical discoveries of yesterday and today, and the cultural voyages of Hawai'i's native people in both ancient and modern times. People representing many views were consulted in planning the MDC, from scientists and observatory staff members to experts in Hawaiian culture to community representatives from Hilo and throughout the state. The facility is designed for everyone, from astronomers and those interested in the mountain's cultural significance to the casual visitor.

The building itself is striking, with three titanium cones representing the Big Island's highest mountain peaks: Mauna Kea, Mauna Loa and Hualālai. Its space-age take on Hawai'i's ancient mountains symbolizes the bridging of science and culture that permeates the center's grounds and internal venues.

The 40,000-square-foot (3,716-square-meter) center includes a planetarium, a large main exhibit hall as well as a smaller one for temporary and traveling exhibits, a gift shop, a restaurant, a classroom, conference and library/resource rooms and more.

Driving into the center, visitors encounter native plant gardens organized to represent plants found in Hawaiian land divisions ranging from the seashore to the summits. Included with these are Canoe Gardens, with plants first brought to Hawai'i in voyaging canoes by early Polynesian seafarers. The center itself, representing the mountain top, has more barren surroundings, characteristic of the high, dry alpine environment that makes Mauna Kea one of the best astronomical viewing sites on Earth.

As you enter the main exhibit hall, Hawaiian and English graphic panels celebrate the renaissance of the Hawaiian language, providing exposure for those interested in the language and practice for those learning Hawaiian. UH Hilo's College of Hawaiian Language, the only university in the nation to offer graduate degrees in an indigenous language, helped to create this bilingual experience.

When you enter the center, you're in the Piko Sense of Wonder orientation area, where visitors are introduced to the connectedness of many Native Hawaiians and astronomers to the mountain. The Forest Walk is a "decompression area" to help you make the transition from outside to the mountain experience. You'll find yourself in a highly stylized representation of mid-level elevation Mauna Kea, specifically the *māmane* forest. Surrounding you are echoes from the past, subtle sounds of Hawaiians harvesting a *koa* tree in order to carve a canoe from its trunk: an occasional command, the crashing of a log.

The Higher Realms area moves you up the mountain to the summit area. There's a stylized representation of Lake Waiau, a recreation of the mountain's adze quarry and a starry sky overhead to demonstrate that people went up the mountain to study the stars. Short narrations about the Mauna Kea deities Līlīnoe and Poli'ahu, the lake, and people working in the adze quarry play in this area. Scattered throughout are graphics of old journal entries by past explorers who visited the summit and wrote about their expeditions.

The exhibits that pertain to the center's Origins theme try to answer the question of where we came from with both cultural and scientific explanations. First you experience the Hawaiian perspective, including an exhibit highlighting the Kumulipo ("Beginning-in-deep-darkness"), a traditional Hawaiian account of the creation of the cosmos similar to the theory of evolution in that both relate the idea of light and life springing forth from darkness.

The 2,000-plus-line creation chant traces the genealogy of a Hawaiian royal family back through its ancestors and then on to great rulers, heroes and gods, as well as to plants and animals, the stars, and the first spark of life in the universe Its English translation begins:

At the time that turned the heat
of the earth
At the time when the heavens
turned and changed
At the time when the light of the
sun was subdued
To cause light to break forth
At the time of the night of
Makalii *[winter]*
Then began the slime which
established the earth,

The source of deepest darkness,
Of the depth of darkness, of the
depth of darkness,
Of the darkness of the sun, in the
depth of night,
It is night,
So was night born.

A sit-down theatrical experience depicts the Kumulipo with a variety of sensory experiences via such media as film and special effects (you actually feel the heat, the mist and the bellowing winds, for instance). You can watch life come about just as it's recounted in the Kumulipo: darkness unfolds, then you see hints of living things before they take shape, all the way through to the birth of Hāloa (the firstborn of the gods Wākea and Pāpa—he was stillborn and from his body grew the first taro) and his brother, also named Hāloa, who was the first man.

The MDC unites the ancient and the cutting-edge for a fun and educational experience.

Exhibits portraying the scientific view of life's origins take a different approach. Step into the Atomic Time Machine and you can explore the history of one of the atoms that makes up your body. Where was that atom before it was a part of you? Was it part of last night's dinner? A dinosaur? A comet? The exhibit traces one of your atoms as it might have existed all the way back to the Big Bang.

Other exhibits explore how life may have begun, how comets might have introduced the building blocks of life, and more. Video animation demonstrates how the solar system came into being, why gassy planets are farther out than others, and many other questions about the universe. An interactive exhibit even lets you combine different elements and see what type of planet they would form.

There are exhibits on the formation and evolution of the moon. See a timeline of moon knowledge—for instance, did you know that in Galileo's time we knew of just five moons and now we know of 170? A great deal of moon research has been done on Mauna Kea. In fact, all throughout the MDC you'll find information on things we now know specifically because of astronomical research done on Mauna Kea.

The center's Voyages exhibits teach about the exploration of both ancient Hawaiians and modern astronomers—though the exploration done by these two groups is somewhat different. Presented side by side are astronomical observatories—ways of exploring other worlds—and Hawaiian voyaging canoes and navigators—vehicles for exploring this one.

The astronomy section will be phasing in a Voyages exhibit that will give you a hands-on experience. There will be full-scale models of

astronomical telescopes to operate and real telescopes to examine. Other exhibits explore ongoing research about such heavy-duty questions as black holes, whether there is life in other worlds, and the ultimate fate of our universe. The Voyaging Through Space Theater, sponsored by the National Astronomical Observatory of Japan, will be a 4D (that's 3D plus time) exploration through the universe, using astronomy photos taken and data collected by today's astronomers. Wearing 3D glasses, visitors will travel through space and experience what astronomers see without leaving the comfort of their seats.

The Hawaiian aspect of the Voyages exhibit starts with a canoe and explains its cultural significance: traditionally, who were voyagers and navigators? How did they make their way across the ocean and why? The exhibit takes the viewer all the way up to the present, a time of a great resurgence of interest in navigating that has led to efforts to recreate the Hawaiian star compass and the teaching of new generations of navigators.

You can learn how Polynesians purposefully and successfully traveled the ocean without the type of instrumentation we take for granted today, and even try your hand at voyaging without instruments in an interactive navigation simulator. See if you have what it takes—can you find Hawai'i?—or if you get hopelessly lost in the Pacific. Or immerse yourself in a modern Hawaiian voyage: experience the preparation needed before starting out, life on board, navigational techniques, battling the elements, sighting land and traditional welcoming ceremonies that are still practiced by many Pacific Islands peoples today. Interactive displays focus on natural and cultural aspects of the mountain, such as how Hawaiians dealt with the extreme cold weather at the high-altitude adze quarry, and the mountain's indigenous plants and their role in Hawaiian culture. One exhibit discusses the connection between the current Hawaiian cultural revival and the revitalization of other indigenous cultures and languages.

The multipurpose planetarium, which seats 124 and is 53 feet (16 meters) in diameter, offers planetarium shows, lectures, conference and video presentations, live performances, live interactions between the center and activities on Mauna Kea, and more. Shows also explain some of the latest discoveries made on the mountain, demonstrate how Hawaiians navigated by the stars and discuss other aspects of astronomy. Different programs are made available to school groups and visitors.

Some of the research happening at the mountain's observatories that you might learn about in your visit to the center includes discovering planets orbiting other stars, mapping the Milky Way Galaxy in the larger universe, discovering that most of the universe's mass may be a mysterious form of dark matter, detecting and identifying organic molecules in comets, and developing new techniques in optics and sophisticated image processing in order to remove "twinkle" from stars (which allows astronomers to get clear and detailed images rivaling those from the Hubble Space Telescope).

Here's another great thing about the center: if you can't visit it in person, or if you find yourself fascinated by what's there and needing to know more, many of its programs in time will be available at *http://manuakea.hawaii.edu* on the Internet. See live images as they are being observed through the Mauna Kea telescopes; read about the significance of discoveries made and other research conducted by astronomers there; learn about the cultural context of Mauna Kea, the latest interpretations of Hawaiian sites on the mountain, developments in cultural revitalization and new connections with other indigenous peoples. Or you can take part in videoconferences on astronomy, Hawaiian and Polynesian navigation and voyaging, or other related topics.

One goal for the MDC is to inspire the next generation of scientists. It also plans to help *train* the next generation of astronomers and space scientists, as well as to improve the level of science literacy in the U.S. To achieve these lofty goals, the MDC welcomes school groups both locally and from around the world. It plans to provide teachers with ongoing professional development by offering local and distance-learning courses and programs, curriculum development, and research opportunities in collaboration with astronomy faculty at the University of Hawai'i at Hilo and Mānoa and at participating observatories.

Additional plans are to offer summer astronomy workshops and courses for different levels of astronomy students, from secondary school through college, utilizing both the center and the Mauna Kea observatories. The center plans to hold year-round workshops on Hawaiian and other indigenous cultures, develops curriculum material on the same, and offers academic programs in the Hawaiian language.

Public lectures at the MDC are conducted by both Mauna Kea astronomers and visiting scientists. Call to see if there are any sky-watching events or other interesting astronomical events on the schedule.

The restaurant, with a panoramic view of Hilo Bay, serves both visitors to the center and staff from neighboring observatories. Have lunch with an astronomer! The gift shop offers gifts, souvenirs and educational items.

When you leave the center, all set to see in person what you've been learning about, turn left on Komohana Street and continue for about 0.4 mile (0.6 kilometer) to the first stoplight. Turn right at this stoplight onto Pū'ainakō Street. After about 4.7 miles (7.5 kilometers), Pū'ainakō Street joins Kaūmana Drive, and after another half-mile (0.8 kilometer), Kaūmana becomes Saddle Road (see p. 11) and takes you to the Mauna Kea turnoff. ▲

The Future 8

8 The Future

FUTURE ERUPTIONS

It's something to ponder when you're about three miles up the mountain: will Mauna Kea erupt again?

Geologists say that it almost certainly will.

You can relax, though, as it's almost definitely not going to blow on the day of your visit. Mauna Kea—which last erupted about 4,400 years ago, long before humans lived on these islands—is considered dormant, and probably will not erupt anytime soon.

Future eruptions will likely be similar to past ones, forming high cinder cones and spewing slow-flowing lava that will mostly impact the lower flanks of the mountain. Since the mountain's glacial period 10,000 years ago, there have been about 12 eruptive events. Eruptions on Mauna Kea are "episodic"—there was a particularly intense concentration of eruptions between 4,400 and 5,600 years ago—and not periodic, so the fact that the volcano hasn't erupted in such a long time isn't a concern. There will be numerous earthquakes before another active period of eruptions happens, probably giving scientists years of warning. There haven't been any such "volcanic earthquakes" below Mauna Kea, the most minute of which would be detected by the sensitive astronomical equipment at the mountain's summit.

So relax and enjoy your visit!

THE MASTER PLAN

The University of Hawai'i's Board of Regents adopted a new Master Plan on June 16, 2000, which will govern the use and development of the Mauna Kea Science Reserve for the next 20 years. Central to this Mauna Kea Science Reserve Master Plan is establishing a mechanism for local control to govern future use and development. To achieve this goal, the Office of Mauna Kea Management (OMKM) was established. This office is vested with the responsibility of implementing the provisions of the Master Plan and reports directly to the Chancellor of the University of Hawai'i at Hilo. The office serves as the single point of contact with comprehensive management authority to protect the sustainability of Mauna Kea's resources. OMKM is also the contact point for the public and functions as a referral and facilitative agency for issues that are outside its authority but related to the mountain. The Mauna Kea Rangers are a component of the OMKM.

Here Mauna Kea casts its shadow onto the atmosphere, creating a ghostly mountain.

The Office of Mauna Kea Management is aided in its mission by the community-based Mauna Kea Management Board, whose members are recommended by the UH Hilo Chancellor and appointed by the Board of Regents. The Mauna Kea Management Board represents the broad Hawai'i community and advises the Chancellor. Several board committees deal with such matters as education, the environment, and health and safety. A special Kahu Ku Mauna ("guardian of the mountain") Council has also been formed, made up of representatives of native Hawaiian organizations and individuals recognized for their specialized knowledge. This council advises the Mauna Kea Management Board and the Chancellor on cultural issues related to Mauna Kea.

The University of Hawai'i Institute for Astronomy (IfA) was founded in 1967. IfA astronomers teach astronomy at the undergraduate and graduate level, conduct astronomical research and manage the astronomical development within the Mauna Kea Science Reserve. Mauna Kea Observatories Support Services (MKSS), a component of the Institute for Astronomy (IfA), is responsible for providing the various support functions for Mauna Kea. These include road maintenance, snow plowing, communications and power systems, managing Hale Pōhaku and the Visitor Information Station, and most other infrastructure needs.

A LOOK AHEAD

The Mauna Kea Science Master Plan allows for several of the old observatories to be upgraded and a few new observatories to be constructed. These new upgrades and observatories must go through a complex approval process specified in the Master Plan to ensure that all stakeholders in Mauna Kea have

Many groups are working together to further develop the summit respectfully.

their concerns heard.

One such project is NASA's proposed Outrigger Telescopes plan, which would entail building, installing and operating four to six "outrigger" telescopes near the existing Keck Observatory. Designed to help answer the questions "Where did we come from?" and "Are we alone?" these proposed 1.8-meter (6-foot) telescopes would be strategically placed around the two 10-meter (33-foot) Keck telescopes to take advantage of interferometry—the combining of light—to function as a single, much larger, telescope, allowing images to be viewed with substantially more detail.

Other planned projects include one by the University of Hawai'i to replace its 0.6-meter telescope with a 1-meter telescope, which would then be operated remotely from the campus as an instructional telescope. This is projected for completion by 2006.

Pan-STARRS is an acronym for the Panoramic Survey Telescope and Rapid Response System. This wide-field imaging system, being developed by the University of Hawai'i's IfA, will detect asteroids and comets that potentially could impact the earth. A major impact occurred about 65 million years ago, leading to the extinction of the dinosaurs. More recently in 1908, a smaller body hit Tunguska, Russia, and caused widespread devastation. Pan-STARRS will consist of four 1.8-meter (6-foot) telescopes, each with a 3-degree field-of-view. The telescopes will observe the entire sky several times per month. Each telescope will be equipped with a billion-pixel imaging array. Both Mauna Kea and Haleakalā are being considered for this project, which will likely be completed in 2008. If Pan-STARRS is built on Mauna Kea, the UH 2.2-meter Telescope will probably be decommissioned and the new telescopes built in its place.

The other big project currently being explored for Mauna Kea is the Thirty-Meter (98-foot) Telescope (TMT). Advancements in telescope technology have made possible the building of an optical/infrared telescope roughly 10 times more powerful than the twin Keck telescopes, currently Mauna Kea's largest. A group of universities and astronomical organizations is considering building the TMT on Mauna Kea; other sites under review are in Chile and Mexico. If it's built in Hawai'i, the first viewing through this telescope would probably take place in 2014. ▲

This time-exposure photo shows Mauna Kea at night, viewed from Mauna Loa. The stars appear to rotate arond the north celestial pole—the point in the sky above the earth's North Pole—because of the earth's rotation.

RESOURCES 9

9 Resources

WEBSITES

Mauna Kea weather and webcams
http://www.maunakea.com/livecams.htm

The Visitor Information Station
http://www.ifa.hawaii.edu/info/vis/

The Institute for Astronomy
http://www.ifa.hawaii.edu/

The Office of Mauna Kea Management
http://www.malamamaunakea.org

The Division of Forestry and Wildlife has a fun kids' page complete with contests, downloadable coloring books and more.

Volcano terms
http://volcanoes.usgs.gov/Products/Pglossary/pglossary.html

A volcano site for kids
http://volcano.und.nodak.edu/vwdocs/kids/kids.htm

Hawai'i State Parks
http://www.hawaii.gov/dlnr/dsp/hawaii.html

Hawai'i Division of Forestry and Wildlife
http://www.dofaw.net

Natural Area Reserve System
http://www.dofaw.net/nars/images.php?reserve=003

Wēkiu
http://www.hcc.hawaii.edu/~pine/Phil100/wekiubug.htm

Mauna Kea Access Road hiking trail
http://www.hawaiitrails.org/trail.asp?TrailID=HA+46+006&island=Hawaii

Hunting information
http://www.state.hi.us/dlnr/dofaw/hunting/hawmammregs.html

William Herschel
http://en.wikipedia.org/wiki/William_Herschel

READING LIST

Beckwith, Martha Warren. *The Kumulipo.* Honolulu: University of Hawai'i Press, 1951. Excellent text that includes the Hawaiian creation chant in Hawaiian, an English translation, and extensive commentary and analysis.

Bryan, E.H., Jr. *Ancient Hawaiian Civilization.* Vermont: Charles E. Tuttle, 1965. Contains useful and interesting information on many aspects of Hawaiian culture, including traditional astronomy.

Bryan, E.H., Jr., and R. Crowe. *Stars Over Hawai'i*, Hilo: Petroglyph Press, 2002. Contains monthly star charts for Hawai'i's skies and basic concepts of modern astronomy.

Cruikshank, Dale P. *Mauna Kea: A Guide to the Upper Slopes and Observatories.* University of Hawai'i, Institute for Astronomy, 1986. The original guidebook to Mauna Kea, this is now dated in terms of the observatories, but the book is still of great interest.

Ellis, William. *Journal of William Ellis: Narrative of a Tour of Hawaii, or Owwhyee: with Remarks on the History, Traditions, Manners, Customs, and Language of the Inhabitants of the Sandwich Islands.* Honolulu: Advertiser Publishing Company, Ltd., 1963. A fascinating and detailed account of an English missionary and his entourage's trek around the island of Hawai'i in the year 1823.

Learn more about Mauna Kea's many exciting phenomena with these resources.

Hazlett, Richard W., and Donald W. Hyndman. *Roadside Geology of Hawai'i.* Missoula, Montana: Mountain Press Publishing Company, 1996. A user-friendly book that divides the islands into areas and explains the geology of each.

Juvik, Sonia P., and James O. Juvik, editors. *Atlas of Hawaii*, 3rd edition. Honolulu: University of Hawai'i Press, 1998. An excellent reference for questions about Mauna Kea and Hawai'i in general.

Kane, Herb Kawainui. *Ancient Hawai'i.* Captain Cook, Hawai'i: The Kawainui Press, 1997. Writer and illustrator Kane has been named a "Living Treasure," and his work is widely admired for its careful attention to detail and accuracy. This book has text and reproductions of paintings about old Hawaiian ways, including those of the adze quarry on Mauna Kea.

Kane, Herb Kawainui. *Voyage, the discovery of Hawaii.* Honolulu: Island Heritage, 1976. A carefully researched, poetic fictional account of an early Hawaiian voyage in the old tradition of navigating by the elements. Out-of-print but available at libraries; used copies are available online.

Morey, Kathy. *Hawaii Trails.* Berkeley: Wilderness Press, 1992. A trail guide to the Big Island with detailed information on the island and its hikes, including several off Saddle Road.

Puku'i, Mary Kawena, Samuel H. Elbert and Esther T. Mo'okini. *Place Names of Hawaii.* Honolulu. University of Hawai'i Press, 1974. Look up Hawaiian place names alphabetically and learn where they are, what they mean, and in many instances how they came to be used.

Rhoads, Samuel E. *The Sky Tonight.* Rev. Ed. Honolulu: Bishop Museum Press, 2000.

Rumford, James. *Dog-of-the-Sea-Waves.* Boston: Houghton Mifflin, 2004. Sequel to *The-Island-below-the-Star,* this story is a fictional account (in both English and Hawaiian) of the same five brothers exploring their new home as the first human residents of the island of Hawai'i. This children's book has beautiful watercolors of the birds, plants and fish that exist only in these islands, each of which are described in end notes.

This 1875 etching showing the British awaiting the Transit of Venus can be found in Walter Steiger's online book Origins of Astronomy in Hawai'i.

An astronomer watches the sun set from the catwalk of the Canada-France-Hawaii Telescope.

Rumford, James. *The-Island-below-the-Star*. Boston: Houghton Mifflin, 1998. An award-winning children's book about five brothers who sail from their Polynesian island to become the first humans on Hawai'i. A fictional account that illustrates methods of Polynesian navigation, including following the stars.

Shallenberger, R.J., editor. *Hawai'i's Birds*. Hawaii Audubon Society, P.O. Box 22832, Honolulu, HI 96822. Various editions. A small, well-respected volume about Hawai'i's birds.

Steiger, W.R. *Origins of Astronomy in Hawai'i*. On the web at *http://www.ifa.hawaii.edu/users/steiger*. The story of how modern astronomy developed in Hawai'i.

Thompson, Vivian L. *Hawaiian Myths of Earth, Sea and Sky*. Honolulu: University of Hawai'i Press, 1966. Thompson tells traditional stories of Hawai'i's natural world, including a legend about Poli'ahu, the snow goddess of Mauna Kea.

Westervelt, W.D. *Hawaiian Legends of Volcanoes*. Rutland, Vermont: C.E. Tuttle Co., 1963. Interesting reading about Hawai'i's traditions surrounding volcanoes.

ACKNOWLEDGMENTS

There are so many we would like to thank. The first acknowledgement should always be to those who came before. The original discoverers of the Hawaiian Islands, and therefore Mauna Kea, sailed their magnificent double-hulled canoes from the Marquesas Islands using the stars to guide them. Later came the Tahitians, again in great double-hulled canoes, using the stars and other way-finding techniques to travel to these beautiful Hawaiian Islands. From these great voyaging peoples came the rich and complex Hawaiian culture and people.

The Hawaiian people and culture gave us the pantheon of gods and goddesses who reside on Mauna Kea. This offering of spirituality is the greatest of gifts. Then, too, we can only marvel at the adze quarry, Keanakāko'i, and wonder at the fortitude and ingenuity of the craftspeople of so long ago who formed these beautiful tools. And finally, they gave us the gift of the beautiful and poetic place names of Mauna Kea in the language with ancient origins that is experiencing a modern rebirth.

Over the years, many people have contributed to the Mauna Kea Observatories. Their names and stories are too numerous to note here, but we who live on, work for and love Mauna Kea can truly say that we stand on the shoulders of giants.

The progress over the years at the Mauna Kea Visitor Information Station is principally due to the perseverance and dedication of Ron Koehler, General Manager of Mauna Kea Support Services. All visitors to Mauna Kea owe a special debt of gratitude to him for his efforts to develop the educational programs and the health and safety protections that we enjoy today.

Bill Stormont, the Director of the Office of Mauna Kea Management, has accepted the challenging task of forming the Office of Mauna Kea Management and implementing the new Master Plan. His leadership and knowledge of the issues related to Mauna Kea and his support of the Mauna Kea VIS and the Mauna Kea Rangers have been critical to all public programs.

Dr. Rose Tseng, Chancellor of the University of Hawai'i at Hilo, provides encouragement to all phases of Mauna Kea's maintenance and development, especially to the students who volunteer and work on Mauna Kea.

Dr. John T. Jefferies, the founding director of the Institute for Astronomy, possessed the foresight and risk-taking ability that provided the leadership to develop Mauna Kea as an international astronomical site of central importance.

Dr. Rolf-Peter Kudritzki, Director of the Institute for Astronomy, has always been supportive of all public-education programs, both scientific and cultural. His vision and multidimensional appreciation of all aspects of Mauna Kea will provide an example for the future of leadership in astronomy.

With the proactive and forward-looking leadership of Dr. Robert McLaren, Associate Director of the IfA, great strides have been made in reconciling the various community issues associated with Mauna Kea. His association with Mauna Kea spans decades.

We cannot sufficiently thank the various components of UH-Hilo and Mānoa. These include the Institute for Astronomy, the Office of Mauna Kea Management, the Physics & Astronomy Department at UHH, Maunakea Discovery Center and Mauna Kea Observatories Support Services. The directors and their staffs have been helpful and supportive in every way.

The Office of Mauna Kea Management and its related committees have always been enthusiastic and supportive of all the efforts of the Mauna Kea VIS and the Rangers. Their staff includes Bill Stormont, Director, Stephanie Nagata, Associate Director, and Dawn Pamarang. Special thanks are due to Kahu Ku Mauna for giving all of us cultural guidance.

The Mauna Kea Observatories not only provide financial support for the free public interpretative programs but also support the various activities in many different ways. We gratefully thank their directors and staffs.

The Mauna Kea Observatories Outreach Committee has provided critical support to all educational and outreach efforts of the VIS. Special thanks to Bobby Bus, Liz Bryson, Richard Chamberlin, Richard Crowe, Gary Fujihara, Catherine Ishida, Ron Koehler, Laura Kraft, Wendy Light, Mike Maberry, Peter Michaud, Janice Harvey, Alison Peck, Douglas Pierce-Price, Antony Schinckel, Paul Coleman, Walter Steiger, Richard Wainscoat and Michael West.

The Astronaut Ellison S. Onizuka Space Center at Keāhole Airport has been helpful with providing information about Ellison S. Onizuka and public programs.

The Hilo Astronomy Club, with its president David Brennen, has provided support and insights into the local amateur astronomy community. We thank them for their many star parties held at the VIS.

The Interpretative Guides, Rangers, student staff and volunteers of the Visitor Information Station consistently provide great, innovative programs and services to all who travel to Mauna Kea.

Thomas Krieger was the General Manager of Mauna Kea Support Services when the Mauna Kea Visitor Information Station was constructed, and he established the initial outreach and interpretative programs of the station. He also set up all the services that Mauna Kea Observatories Support Services provides to the observatories. Gail Forbes, Administrative Officer of Mauna Kea Support Services, has always been supportive of station operations and programs. During the early years of station operations, only one part-time interpretative guide ran all of the VIS programs. These pioneers were Donald Romero, Su Reed, Tom Peek, Donald Burciaga and Hugh Grossman.

Many thanks to the Interpretative Guides: Gary Beals, Kenyan Beals, Rich Berner, Koa Ell, Erik Rau and Shane Fox.

The Rangers have been helpful in many ways. Mauna Kea would be a poorer place without them. All who travel to Mauna Kea owe them a debt of gratitude. Many thanks are due to Rangers Trevor Anderson, Kenyon Beals, Ahiena Kanahele, Pablo McLoud and Kimo Pihana.

The student staff of the VIS provides the energy and enthusiasm that only youth can bring. There have been so many students over the years that it would be impossible to list them all. Their contributions to the educational programs of the VIS will be remembered and appreciated by all they have touched.

The Volunteer Corps is one of the treasures of Mauna Kea. Between July 1, 2003, and June 30, 2004, 133 volunteers contributed 9,487 hours. Words cannot express the impact these wonderful volunteers have on everyone who travels to Mauna Kea. Their contributions to educational programs, support of the Rangers, and trail maintenance have made a significant quality difference in all areas. "Volunteer of the Year" awardees are Richard A. Berner—2000, Alexandre Y. K. Bouquin—2001, Jonn L. Altonn—2002, Alanna A. Garay—2003 and Marc Seigar—2004. These individuals represent all the precious volunteers who are too numerous to name here.

The following people reviewed the text: our many thanks for your time and efforts. Paul Coleman, Mike Maberry, Douglas Pierce-Price, Antony Schinckel, Alison Peck, Catherine Ishida, Laura Kraft, Bobby Bus, Richard Crowe, Gary Fujihara, Peter Michaud, Ron Koehler, Arnold Hiura, Robert McLaren, Liz Bryson, Mark McKinnon, Marlene Hapai, Jim Kanahikawa, Gregory Brenner, Rob Pacheco, Walt Steiger, Gary Beals, Pat McCoy, Holly McEldowney, Kimo Pihana, Richard Wainscoat, Mercedes Stevens, Christian Veillet, Lyman Perry, Ryan Lyman, Trevor Anderson, Rich Berner, Koa Ell, Eric Rau, Justin Stevick and Shane Fox all contributed their knowledge and time. Thanks also to Frederika Bain of Watermark Publishing for helping coordinate the book's production and extensive review process.

Many others helped in many different ways. To those we may have forgotten, please accept our apologies—and our sincere thanks for your contributions.

—Dave Byrne
and Leslie Lang

Special thanks to my dear sweet wife Heidi, who so patiently and lovingly tolerated my many weeks away from home. Heidi spent many hours proofing various sections of this book. My niece, Rebecca Lynn Ford, also helped with proofing the text.

—Dave Byrne

Credits

Unless otherwise attributed, all photos and maps courtesy VIS. Mahalo to the station and to the many visitors who shared their photos.

Margaret Barnaby, 60

Alex Bouquin and Roerto Avila, 110

Caltech Submillimeter Observatory, 98

Jean-Charles Cuillandre (CFHT), © 1999, 42-43, 44, 45, 47, 84-85, 119, 128-129, 131, back cover

Bill de Ment, 122

DLNR-Forestry and Wildlife and Mililani Mauka Elementary School, bottom 136

Gemini Observatory/AURA, 117

Hawai'i State Archives, 11, 39, 40, top 59, 68

G. Brad Lewis; permission courtesy Photo Resource of Hawaii, 62-63

Macario, 22-23. Of his photograph, Macario writes, "Kahuna Nui Frank Pao, of the Royal Order of Kamehameha, with Na Koa Keoni Choy and Harold Kaula during a Winter Solstice ceremony on Mauna Kea."

Barney Magrath, 16, 86

Peter Michaud, 133

Maunakea Discovery Center, 120-121, 124-125

National Radio Astronomy Observatory/Associated Universities, Inc./ National Science Foundation, 96, 97

C. Sanchez; photo Hugh R. Grossman, © 1998, 58

S.C. Smith; photo Hugh R. Grossman, © 1998, 29

Walter Steiger, 138

Subaru Telescope, NAOJ. All rights reserved, 104, 105

Submillimeter Array, 102

United Kingdom Infrared Telescope, 112

Richard J. Wainscoat, front cover, ii, 70-71, 109, 111, 115, 134-135, 139. Of his photograph on pp 134-135, Dr. Wainscoat writes, "The full Moon is seen rising inside the shadow of Mauna Kea on the atomosphere. Although the shadow of Mauna Kea can be seen most days at sunrise and sunset, it is very rare to be able to see the Moon inside the shadow, because the Sun, Earth and Moon must be precisely aligned." All photos © 2000, all rights reserved.

Wei-Hao Wang, 114

W. M. Keck Observatory, 106

INDEX

About the Authors

DAVID A. BYRNE

David A. Byrne holds a BS in Geology & Geophysics and a Master's degree in Business Administration from the University of Hawai'i. He accepted the position of Mauna Kea Visitor Information Manager in 1998 and has expanded the operations and programs at the Station to make it the world-class facility it is today. David has many scientific and popular publications to his credit. In addition to his long association with Mauna Kea, he has spent three years aboard various geophysical research vessels and has led a design group that developed a new class of geophysical instrumentation. During his tenure at the Hawai'i Institute of Geophysics, he was also a Dive Team Captain. He has held positions with the Hawai'i Institute of Geophysics, the Salvation Army, the Waikīkī Aquarium and the Bishop Museum. When not working on Mauna Kea, David takes Hawaiian language and culture courses at the University of Hawai'i at Hilo and is building a new home at Volcano.

LESLIE LANG

Leslie Lang is a freelance writer and editor whose work has appeared in newspapers, regional and national magazines and literary journals, as well as on public radio. Her educational background includes a BA in journalism and an MA in anthropology with an emphasis on cultural Hawaiian and Pacific anthropology. She frequently writes about Hawaiian culture, both ancient and modern; she also writes other features, travel articles, essays and more. She has been awarded an "Excellence in Journalism" award by the Society of Professional Journalists, and two of her articles won "Keep It Hawai'i" awards from the Hawai'i Visitor and Convention Bureau. She lives and writes on the flanks of Mauna Kea, in Hāmākua, with her husband and daughter.